Fuzzy Management Methods

With today's information overload, it has become increasingly difficult to analyze the huge amounts of data and to generate appropriate management decisions. Furthermore, the data are often imprecise and will include both quantitative and qualitative elements. For these reasons it is important to extend traditional decision making processes by adding intuitive reasoning, human subjectivity and imprecision. To deal with uncertainty, vagueness, and imprecision, Lotfi A. Zadeh introduced fuzzy sets and fuzzy logic. In this book series "Fuzzy Management Methods" fuzzy logic is applied to extend portfolio analysis, scoring methods, customer relationship management, performance measurement, web reputation, web analytics and controlling, community marketing and other business domains to improve managerial decisions. Thus, fuzzy logic can be seen as a management method where appropriate concepts, software tools and languages build a powerful instrument for analyzing and controlling the business.

More information about this series at http://www.springer.com/series/11223

Zoumpolia Dikopoulou

Modeling and Simulating Complex Business Perceptions

Using Graphical Models and Fuzzy Cognitive Maps

Springer

Zoumpolia Dikopoulou
Faculty of Business Economics
Hasselt University
Hasselt, Belgium

ISSN 2196-4130 ISSN 2196-4149 (electronic)
Fuzzy Management Methods
ISBN 978-3-030-81495-3 ISBN 978-3-030-81496-0 (eBook)
https://doi.org/10.1007/978-3-030-81496-0

This Springer imprint is published by the registered company Springer Nature Switzerland AG.
The registered company address is: Gewerbestrasse 11, 6330 Cham, Switzerland

To a new life…

Foreword

Data science and big data are today very popular terms used in academic and business media today. The research described in the monograph fits in this context. Applying correctly statistical or data mining methods is obviously an essential and trivial element of a data science PhD project. In this research with a huge volume of perception data, knowing, using, and implementing the most appropriate methods was a big challenge and was not trivial at all. The author used graphical models and fuzzy cognitive maps. Fuzzy cognitive maps are able to represent vague and complex scenarios, and they are capable of efficiently solving numerous problems across a variety of domains such as decision support, system control, time series forecasting, pattern classification, and many others. However, there is still a need for robust data-driven models. The general scope of this monograph is to provide a framework for constructing fully data-driven fuzzy cognitive maps based on huge cross-sectional ordinal data. The developed publicly available R-package is a highly valuated extra achievement. These publications (monograph and software) are more than welcome today.

Big data combined with analytics has always one serious challenge: Even the simplest method of analysis can produce an overwhelming number of results. So, data scientists have to prioritize what is important and what is not and present only relevant material. So, selecting and filtering results and choosing the right presentation to communicate, whether they be graphs, simulations, plots, tables, charts, or text, is very important. The author of this monograph Dr. Zoumpolia Dikopoulou does this job in an excellent way. Dr. Zoumpolia Dikopoulou describes perfectly how the business user should "listen to the data" and should obtain clear insights by running simulations.

Working together with Dr. Zoumpolia Dikopoulou was and still is a real pleasure for me. I hope that everyone reading and studying the material of the monograph may join this pleasure and can make forward steps in the reader's research or application domain.

Research group of Business Informatics,
Faculty of Business Economics
Hasselt University
Diepenbeek, Belgium
December 2020

Koen Vanhoof

Acknowledgments

It is hard to believe that this journey ends. In fact, it was a bitter-sweet journey with all the colors of emotions. However, I never regret this decision because this procedure made me more mature since I had the opportunity to work in an academic environment, extend my knowledge to the scientific field of computer science, attend various international conferences and summer schools, and live abroad.

Firstly, I would like to express my deepest appreciation to my supervisor Prof. Dr. Koen Vanhoof for his consistent guidance, support, encouragement, and patience throughout this research project. I remembered the first day we met in Greece. You were quiet and kind (as usual). Nevertheless, after all these years, I can admit that you are not only an inspired professor, a mentor, and an intelligent businessman but also a great person! What I admire from your personality is that you give space and promote all of your PhD students. I am so honored to be your academic "child." I am extremely grateful to my co-promoter Prof. Dr. Bart Kuijpers for his dedicated time to review this dissertation and his valuable advice, suggestions, and comments. I am also deeply thankful to my co-promoter Prof. Dr. Elpiniki Papageorgiou for introducing me to the interesting academic world and teaching me how to stand on my feet. Moreover, I would like to express my deepest appreciation to my committee for their fruitful comments on the dissertation. I would like to extend my sincere thanks to Prof. Dr. Dimitris Karlis for our great collaboration. Because of you, I learned that I must never underestimate statistics, deliver "clean" work, and always trust the resulting numbers (not the visualizations). I would also like to extend my sincere thanks to Prof. Dr. Edy Portman for the opportunity to publish my thesis and for his valuable comments and his practical suggestions. I admire your passion to work and I really enjoyed our relaxed conversations in Fribourg and Naples. You are a person that you know what you want and where to find it. Certainly, you have a business mind! I would also like to thank Prof. Dr. Leticia Arco for her professionalism and her kindness. I hope to meet you again soon and meet your baby as well.

I would also like to gratefully acknowledge and thank my officemates (Farnaz, Aziz, Hamzah, and Ahmad) who have a special place in my heart. Farnaz and Aziz, I feel so fortunate that I met you. You are wonderful people and professionals. I cannot forget how helpful and supportive you were to me. Hamzah and Ahmad, thank you for the amazing atmosphere in which we keep discovering new things. You knew how to cheer me up. Hamzah, I enjoyed our coffee breaks and lunchtime at the university.

My heartfelt appreciation goes to Roberto, an amazing person who changed my life. I feel more relaxed and happier when I am with you because you are generous, careful, protective, and generally, a perfect partner to me. Last but never least, I would like to express the deepest appreciation to my family for their support and encouragement throughout my study. However, mum and dad, I cannot promise that I will stop reading; now the damage has been done. Sotiri (my brother), probably you do not have any idea how much you helped me with your hilarious expressions. I truly need them in the "dark" moments.

August 2020 Zoumpolia Dikopoulou

Contents

1 Introduction 1
1.1 What Is Job Satisfaction? 1
1.2 Model Structures from Data 2
1.3 The Simulation Process for Decision Making 4
1.4 Scope and Research Objectives 5
1.5 Thesis Outline 6
References 10

2 Data Analysis 13
2.1 Introduction 13
2.2 Data Description 14
2.3 Data Pre-processing 15
2.4 Data Visualizations 16
2.5 Conclusions 23
References 24

3 Fuzzy Cognitive Maps 27
3.1 Introduction 27
3.2 Fuzzy Cognitive Maps (FCMs) 30
3.2.1 The History of the Fuzzy Cognitive Maps 30
3.2.2 FCM Inference and Simulation 32
3.3 The 'fcm' R Package for FCM Inference Process 34
3.4 Recommendation to the 'fcm' Users 40
3.5 Conclusions 40
References 41

4 Data Modeling 43
4.1 Introduction 43
4.2 Classification of Networks 45
4.3 Modelling Networks from Cross-Sectional Data 47

4.3.1 The Distance-Based Method . . . 47
4.3.2 Model Selection Using Graphical Models . . . 49
4.4 Modelling on Business Network Structures . . . 57
4.4.1 Results of the Distance-Based Method . . . 62
4.4.2 Results of Graphical Lasso . . . 65
4.5 Comparisons of the Two Methods . . . 72
4.6 Conclusions . . . 79
References . . . 80

5 Network Analysis, Accuracy and Stability of the Job-Satisfaction Structures . . . 85
5.1 Introduction . . . 85
5.2 Centrality Measurement . . . 86
5.2.1 Network Inference . . . 87
5.2.2 Clustering Coefficient and Transitivity . . . 90
5.2.3 Average Path Length . . . 91
5.2.4 Small-Worldness . . . 92
5.3 Network Accuracy . . . 93
5.4 Network Centrality Stability . . . 94
5.5 Network Analysis in Business . . . 95
5.5.1 Centrality Analysis in a Job-Attractiveness Problem . . . 96
5.5.2 Cluster Coefficients in a Job-Attractiveness Problem . . . 100
5.5.3 Associating Clustering Coefficients and Centrality . . . 101
5.5.4 Small-Worldness in Job-Attractiveness Networks . . . 103
5.6 Network Accuracy in Business Networks . . . 106
5.7 Network Stability of Job-Satisfaction Networks . . . 114
5.8 Conclusions . . . 116
References . . . 116

6 The Proposed Data-Driven glassoFCM Method . . . 119
6.1 Introduction . . . 120
6.2 The Graphical Lasso FCM Algorithm (glassoFCM) . . . 122
6.3 Evaluating the Network Stability of the New Job-Satisfaction Models . . . 123
6.4 The Results of the glassoFCM in Job-Satisfaction Problem . . . 124
6.4.1 The Significant Variables in the FCM Structures . . . 128
6.4.2 Statistical Comparisons of the FCM Simulations . . . 130
6.5 Conclusions . . . 133
References . . . 134

7 Thesis Conclusions . . . 135
7.1 Introduction . . . 135
7.2 Research Conclusions . . . 136
7.3 Contribution to Knowledge . . . 138

7.4 Recommendations for Practitioners 141
7.5 Limitations 141
7.6 Future Research 144
References 144

Appendix A 147

Appendix B 149

Appendix C 153

List of Abbreviations

3PLs	Third-party logistics providers
AHP	Analytic hierarchy process
ANOVA	Analysis of variance
BIC	Bayesian information criterion
BPM	Business process modeling
BPR	Business progress reengineering
CI	Confidence interval
CPU	Central processing unit
CRAN	Comprehensive R Archive Network
CS	Correlation stability
EBIC	Extended Bayesian information criterion
ECM	Enterprise content management
EFA	Exploratory factor analysis
ER	Erdős–Rényi
FA	Factor analysis
FCM	Fuzzy cognitive map
FIM	Fuzzified influence matrix
GGM	Gaussian graphical models
GIGO	Garbage in Garbage out
glasso	Graphical least absolute shrinkage and selection operator
glassoFCM	Graphical least absolute shrinkage and selection operator and fuzzy cognitive map
IIM	Initial influence matrix
IT	Information technology
IVM	Iterative vector matrix
MRF	Markov random fields
PDC	Performance-driven change
RAM	Random access memory
SAD	Sum of the absolute differences

SD	Standard deviation
SEM	Structural equation modeling
SIRM	Strength of influence relationships matrix
SISP	Strategic planning of information systems
SW	Small-worldness
URL	Uniform resource locator
WS	Watts and Strogatz

List of Figures

Fig. 1.1 The total number of publications searching for the keywords "job satisfaction" in Scopus. The publications are gradually increasing over time; specifically, from 2004 to 2019, more than 2000 studies yearly are recorded 2

Fig. 1.2 *Structural Equation Modelling* (SEM; left): Circles represent the Latent variables and squares manifest the items of the SEM. *Network structure* (right): Circles depict the observed variables. Moreover, positive and negative relations are represented with blue and red arrow edges, respectively 3

Fig. 1.3 The highlights of each chapter. **RQ** denotes the *Research Question* as described in *Scope and Research Objectives* 7

Fig. 2.1 The procedure of assessment of the job-satisfaction variables for specific companies. Then, follows the cleaning procedure, the filtering process and finally, the required visualizations 16

Fig. 2.2 The demographic information (gender, age, activity and education) of the clean dataset, including 54,246 employees who work in 15 working sectors in Belgium 17

Fig. 2.3 The number of observations of the 15 working sectors ($S_1 - S_{15}$) in Belgium 17

Fig. 2.4 A total of 56,246 respondents replied to the survey concerning job satisfaction. Each employee evaluated one of five levels of agreement or disagreement of 11 variables 18

Fig. 2.5 The Likert-type scales of the ordinal responses in the feature V_{11} (Willingness to work) in every sector. It is observed that no one of the employees responds to this question with the "Strongly Disagree" scale; nevertheless, the negative effect is visible 19

Fig. 2.6 The visualization of the correlation matrix of 56,246 observations to show the associations among 11 variables. Blue and saturated circles indicate positive and stronger associations 21

Fig. 2.7 Low and medium positive bivariate correlations ($0.26 \leq r_s \leq 0.64$) are identified from the dataset of 56,246 observations. The blue color of the lines indicates positive associations; while, the strength of the connections is visualized with thick lines for strong associations and with less saturated lines for weaker connections ... 21
Fig. 2.8 Plot visualization of Spearman associations between two variables in every sector. Rows represent the pairs of variables and columns depict the working sectors ... 22
Fig. 3.1 The construction of an FCM in four steps. This FCM includes four concepts and six causal relationships. Blue and red edges denote positive and negative causal relationships, respectively; while, the saturated and wider edges indicate stronger relationships ... 32
Fig. 3.2 The inputs, the settings and the output of the fcm.infer ... 35
Fig. 3.3 The fcm.infer function to simulate the observed job-satisfaction network. The `fcm.infer()` function requires seven arguments of the: activation vector, weight matrix, iteration, inference function, transformation (threshold) function, lambda and residual ε ... 38
Fig. 3.4 The R environment. The source panel (up left) presents the values of the concepts in every iteration step. The console panel (down left) inform the user in which step the system reaches the equilibrium point and, in the plot/help panel (down right), the plot of the converged values is depicted. The horizontal axis (X-axis) represents the iterations of the FCM simulation; while, the vertical axis (Y-axis) points out the values of the concepts at every simulation step ... 39
Fig. 4.1 A network of five nodes and six edges. Each edge carries a weight indicating the strength of influence among nodes. Blue edges specify positive relations; while, red edges denote negative relations between nodes. (**a**) illustrates an undirected weighted network (left) and (**b**) illustrates a directed weighted network (right) ... 46
Fig. 4.2 The algorithm of the regularized graphical model ... 56
Fig. 4.3 Visualization of the domain of the standard normal density function discretized by the vertical lines which represent the thresholds ... 57
Fig. 4.4 The plot shows the number of methods (in yellow), the eigenvalues (red line) and the cumulative proportion of explained variance (blue line) of the Automotive sector ... 59
Fig. 4.5 The job-satisfaction variables of the Automotive sector (S_1) applying the *k*-means clustering (*k*) of 2, 3, 4 and 5 clusters ... 60

Fig. 4.6 Visualization of three clusters of ten observed job-satisfaction variables using *k*-means clustering method. Each Sector belongs to one of the four groups (**a–d**) 61
Fig. 4.7 Graph visualization of distance-based method applied in Automotive dataset. The derived network is considered as a complete graph since $\delta = 1$. The nodes are colored according to the outcome of the *k*-means clustering method (Fig. 4.5) 65
Fig. 4.8 Visualizations of complete structures ($\delta = 1$) of four sectors using the distance-based method. The nodes of these networks are colored according to the outcome of the *k*-means clustering method (Figs. 4.5 and 4.6) 66
Fig. 4.9 Graph visualization of graphical lasso method applied in Automotive dataset. Blue and red edges represent positive and negative associations, respectively. Wider and more saturated edges indicate higher strengths. The colors of the nodes are depending on the *k*-means results (Fig. 4.5) 67
Fig. 4.10 Graph visualization of graphical lasso method applied in four sectors. The connectivity of the estimated networks ($\delta < 0.89$) indicates sparser structures comparing to complete structures ($\delta = 1$). The nodes are colored according to the outcome of the *k*-means clustering method (Figs. 4.5 and 4.6) 69
Fig. 4.11 A Correlogram which visualizes the partial correlation coefficients (weights) with EBIC regularization among two variables after conditioning with the remaining observed variables in every sector. Blue color represents positive weights, red color indicates negative weights, while white color denotes the absence of partial correlation between the pair of variables after conditioning on all the other variables in the network 71
Fig. 4.12 The Normal Q-Q Plots examine the normal distribution of the weight coefficients of the (**a**) distance-based and the (**b**) graphical-lasso methods in the Automotive sector (S_1), respectively. The dashes straight line denotes the normal distribution in the Q-Q plots 76
Fig. 4.13 Distributions of 15 sectors indicate that the weights derived from the distance-based are normally distributed; while, in Graphical Models, all distributions are left-skewed 77
Fig. 4.14 The estimation of elapsed time in seconds of the distance-based and the graphical lasso method 78

Fig. 5.1 Three basic network types in the model of WS (Watts & Strogatz, 1998). Each graph consists of 20 nodes (N = 20). At the left-side graph, each node has four neighbors (k = 4). Thus, it is defined as a high clustering coefficient C_{cl} and a high average path length L (p' = 0). Increasing the rewiring probability p indicates that randomly edges are rewired and reconnected to arbitrary nodes. Rewiring all nodes of the regular lattice network (p' = 1), the graph becomes completely random (right-side), and properties such as low clustering coefficient and low average path length are showing up. If p' value is between 0 and 1, $0 < p' < 1$, a small-world network is emerged, which incorporates the high clustering coefficient of regular networks with the low average path length of random networks 93

Fig. 5.2 Representation of the job satisfaction graph of Retail section (top left). The color of the nodes denotes the classification in one of the three clusters; Future (green), Stability (blue) and Comfort (pink). Figure 5.2b reports the values of degree, strength, closeness and betweenness indices of ten nodes (top right). The degree distribution shows the spread of nodes' degrees in the network (bottom right) and the visualization of the raw centrality measures for each node is demonstrated at the bottom left 96

Fig. 5.3 The importance of each node of the job-attractiveness networks quantified in centrality indices. Centrality indices (strength, closeness and betweenness) have been standardized (z-scores) to facilitate the comparisons between the job satisfaction networks ... 97

Fig. 5.4 The clustering coefficient of each node of 15 job-attractiveness networks which are separated into four groups according to K-means. As the local clustering coefficient is closer to one, indicates the redundancy of the node in its neighborhood 101

Fig. 5.5 Combining centrality measurements and clustering coefficients retail working sector (S_{10}). The most central nodes (V_2, V_4 and V_9) do not present high clustering coefficients. The values of closeness are multiplied by 100 102

Fig. 5.6 Normal Q-Q plots and degree distributions of the least normal distributed strength-centrality (**a** and **b**) of the S_{10} dataset and the highest normal distributed strength-centrality (**c** and **d**) of the S_6 dataset 105

Fig. 5.7 Bootstrapped confidence intervals for the estimated job-satisfaction network in the Energy sector 107

Fig. 5.8 Bootstrapped confidence intervals of estimated for the estimated job-satisfaction network in 14 working sectors 108

Fig. 5.9 Bootstrapped quantile intervals of each pair of weights in Energy Services (S_{10}). The values in the box indicate the percentage of bootstrapping to be zero. The transparency of the intervals signifies how often an edge is estimated in the bootstrap networks. For instance, the edge V_4–V_8 was estimated as zero in 1320 out of 1500 networks (88%) ... 112
Fig. 5.10 Bootstrapped quantile intervals of each pair of weights in Informatics sector. The bootstrap means and the sample points are very close reflecting an adequate accuracy for the network ... 113
Fig. 5.11 Simulation results using the case-dropping bootstrap in four sectors. Every panel displays the average correlations between centrality indices of the original job-satisfaction networks with the centrality indices obtained from the datasets. Lines represent the average of centrality indices and the colored areas specify the range of the corresponded centrality indices from the 2.5th to the 97.5th quantile ... 115
Fig. 6.1 The data-driven glassoFCM algorithm which models and simulated cross-sectional datasets without any expert knowledge or historical data ... 122
Fig. 6.2 The procedure to obtain FCM simulations from graphical lasso models (steps 1–10). Moreover, the output values of V_{11} are compared to validate if they are statistically significantly different in the observed cases (step 11) ... 125
Fig. 6.3 The pre-processing of the Business sector to identify the initial vectors and the initial matrix for the FCM simulation ... 127
Fig. 7.1 The simulation of *fully-automated* network structures (data-driven model; left) and the simulation of *semi-automated* graph structures (expert-based model; right) ... 142
Fig. B.1 Fifteen corrplots that represent the Spearman's correlations among ten variables in each working sector. Blue color signifies positive correlations among variables ... 149

List of Tables

Table 2.1	The employee satisfaction questionnaire is formatted in a five-point Likert scale. The first ten variables (V_1–V_{10}) reflect the input features correlated to the job-satisfaction problem and the last variable (V_{11}) indicates the output (the decision) variable ...	15
Table 3.1	The result of the `View()` function. The weights of the lower triangular matrix are set equal to zero ...	37
Table 4.1	(a) Reports the symmetric adjacency matrix (left) for the undirected network and (b) displays the asymmetric adjacency matrix (right), respectively ...	45
Table 4.2	(a) Represents the symmetric weight matrix (left) for the undirected network and (b) displays the asymmetric weight matrix (right) for the directed network of Fig. 4.1a, b, respectively ...	46
Table 4.3	The Initial Matrix of the Automotive Sector that is used to calculate the network applying the distance-based method ...	63
Table 4.4	The Fuzzified Influence Matrix (FIM) of the distance-based method of the Automotive Sector ...	63
Table 4.5	The SIRM matrix of ten variables correlated to a job satisfaction problem of the Automotive sector (S_1) ...	64
Table 4.6	The weight adjacency matrix of ten variables using the graphical lasso in the Automotive sector (S_1). Moreover, 17 out of 45 connections are distinguished as spurious (the weights are estimated equally to zero) ...	68
Table 4.7	The actual number of edges and the corresponding density of each graph for every sector estimated by the distance-based and graphical lasso models ...	70
Table 4.8	Measurements of descriptive statistics, a test of normality and Mann-Whitney (U test) to identify the significance of difference among two observed methods in 15 sectors. DB indicates distance-based and GL signifies graphical lasso method ...	74

Table 5.1 Pearson's correlations of node centralities and clustering coefficients of 15 sectors. Strength is correlated strongly and positively with Closeness and with Betweenness Centrality; likewise, the association between Betweenness and Closeness is positive and high. On the contrary, the Clustering coefficient is negatively associated with centrality measurements in the majority of the sectors 100

Table 5.2 Descriptive statistics and test of normality of (N = 10) strength-centrality values of 15 sectors. Observing the p-values of the Shapiro-Wilk test, the strength of every sector follows a normal distribution, except for S_{10} ($p < 0.05$) 104

Table 5.3 According to small-worldness index of each working network, none of the networks holds the small-world property since $SW < 3$. Moreover, the values of Transitivity (T) and Average Path Length ($\bar{L}$) are similar to the estimated T_{est} and L_{est} of random graphs. Consequently, these job-attractiveness networks are classified as pseudo-random 106

Table 5.4 The sum of the absolute differences (SAD) among the sample weights and the mean bootstrap weights. Lower SAD values are observed in larger datasets (S_8 and S_{15}) indicating more accurate networks since the differences between the true model and the mean of the 1500 bootstrap models is very low ($SAD \leq 1$) while higher SAD values are detected in smaller datasets (such as in Energy sector, S_{10}) 111

Table 5.5 Maximum drop proportions to retain correlations of 0.7 in at least 95% of the samples of centrality indices, CS(cor = 0.7) ... 115

Table 6.1 Maximum drop proportions to retain correlations of 0.7 in at least 95% of the samples of centrality indices, CS(cor = 0.7) in job-satisfaction networks of 11 variables. Automotive (S_1) and Energy services (S_{10}) are removed from this study since they are unstable 124

Table 6.2 A network analysis of three hierarchical directed FCMs 126

Table 6.3 The simulation of the V_{11} values for the six observed cases in the Business sector 128

Table 6.4 The comparisons of the output values (V_{11}) to determine which of the observed activations achieves the highest output value. The variables are initiated according to a) the neighbors of the output variable and b) the highest mean variables. The validations are accomplished by filtering the initial dataset 129

Table 6.5 Normality Test of three sectors. As the Sig. value (p-value) under the Shapiro-Wilk column is greater than 0.05, it is concluded that the observations of the correspondent case are normally distributed 130

Table 6.6 The output of the ANOVA analysis of (a) Finance and (b) Non-Profit show that the differences in the cases are not significant. Additionally, a Kruskal-Wallis H test determined that there was a statistically significant difference since $p > 0.05$ 131
Table 6.7 Test statistics table. Tukey multiple comparisons test is applied when data are normally distributed (Finance and Non-profit sectors) and the Mann-Whitney test is used when data significantly deviate from normality (Business sector) 132
Table A.1 Demographic information about the number of observations, gender, education level and working activity of 15 sectors 148
Table C.1 The SIRM matrix (distance-based method) of 11 variables correlated to a job satisfaction problem of the Business sector (S_{12}) dataset. The first ten variables are considered as input variables (V_1–V_{10}) and the last one as the output variable (V_{11}) 154
Table C.2 The 15 FCM simulations of the V_{11} values for the six observed cases in the Business sector. The upper-triangular FCM weighted matrix is derived from the distance-based method (Table C.1) 154

Chapter 1
Introduction

This chapter introduces the job satisfaction problem, why it is important, when the term job satisfaction appeared for the first time in the literature, what are the relationships between employer and employee, what are the benefits of job satisfaction to the employee but also to the company etc. Moreover, terms such as data modeling, simulation and decision-making are familiarized. Finally, the scope, the research objectives and the outline of this thesis are described.

1.1 What Is Job Satisfaction?

Job satisfaction is the evaluation of various concepts that can describe how employees think about their job (Weiss & Merlo, 2015). In the business world, managers are concerned about the employees' job satisfaction since it is important for business attractiveness (Spector, 1997b). Frequently, this topic is studied in organizational behavior research, while the process to measure the aspects of job satisfaction can be accomplished through interviews or questionnaires (Spector, 1997a). First Hoppock in 1937, introduced the term of job satisfaction in which 203 responses of members and affiliates of the American Psychological Association were evaluated (Hoppock, 1937). According to Scopus results, from 1937 until today, the popularity of job satisfaction in the last 20 years has been increased as thousands of studies have been performed (Fig. 1.1).

Moreover, psychologists have categorized the job satisfaction in two groups concerning: i) the overall satisfaction in which the job is evaluated as a whole and ii) the facets satisfaction in which different features of the job satisfaction such as salary, job environment, etc. are evaluated individually (Weiss & Merlo, 2015). According to Tang et al. (2014), three main relationships affect the employee satisfaction: i) the employee—organization relationship which includes other aspects such as the employee's work-family balance, ii) the employee—supervisor relationship and iii) the employee—co-worker relationship.

Z. Dikopoulou, *Modeling and Simulating Complex Business Perceptions*, Fuzzy Management Methods, https://doi.org/10.1007/978-3-030-81496-0_1

Fig. 1.1 The total number of publications searching for the keywords "job satisfaction" in Scopus. The publications are gradually increasing over time; specifically, from 2004 to 2019, more than 2000 studies yearly are recorded

Furthermore, job satisfaction is a multifaceted concept which can reward the employees and the employers. For employees, many features can be improved like salary, job stability, career growth, etc. For employers, a satisfied employee always contributes more to a company. Generally, a happy employee is a productive employee since job satisfaction affects positively the job performance (Locke, 1970). Moreover, the significant challenges of employers focusing on: i) the employees' needs, ii) the improvement of job satisfaction aspects and iii) fitting the needs of the employees with the goals of the company. Improving job satisfaction features, more employees will be satisfied; therefore, their loyalty will be increased and workers will be 'ambassadors' for the company. Furthermore, recruitment costs and employees' training will be reduced, while the HR (Human Resources) will be able to recruit talented workers and save money.

1.2 Model Structures from Data

As data modeling is referred to the process of creating a model from a dataset that can be useful for recognizing patterns, making decisions, predictions and control systems (Bruton & Beyeler, 2019). In other words, data modeling is a visual representation of data which it is regularly visualized as a graph. Fundamentally, a graph is a very simple notion. It is composed of two important elements or vertices and edges (Newman, 2010). In computer science, these are known as nodes and

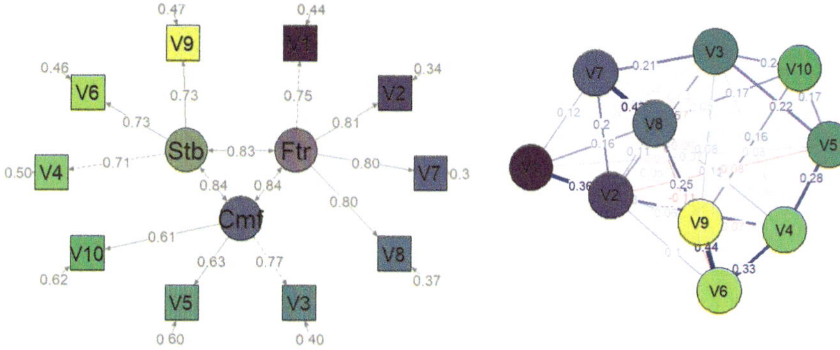

Fig. 1.2 *Structural Equation Modelling* (SEM; left): Circles represent the Latent variables and squares manifest the items of the SEM. *Network structure* (right): Circles depict the observed variables. Moreover, positive and negative relations are represented with blue and red arrow edges, respectively

links, respectively. Each node of the graph represents a variable (feature or item) of the dataset, while each link signifies the association between two observed variables.

There are many methods to estimate a model from a single dataset. The choice of models depends on the variety of variables, the sample size of data and the available knowledge (if any) from experts. Hence, the selection of a model could be the one that best fits into the dataset (Brunton & Beyeler, 2019). Fitting expert knowledge requires relatively less data and leads to simpler interpretable model. On the other hand, data-driven models require a sufficient amount of training data; however, they are more expressive since they can uncover unexpected patterns of the data comparing to expert-based models.

Over recent years, researchers investigate methods to model data-driven problems from cross-sectional data as a causal view of latent factors known as Latent Space Models (LSMs; Brunton & Beyeler, 2019). Factor Analysis (FA) belongs to a family of LSMs in which measurable variables are represented from unobserved latent variables or hidden causes where $r < < p$ (Thompson, 2007; Thurstone, 1931). In other words, the goal of factor analysis is to model the interrelationships between items with fewer (latent) variables by sharing the amount of variance among a set of items and the latent variable. For the record, Structural Equation Modelling (SEM) is a combined method of factor analysis and multiple regression analysis which it examines and visualizes the structural relationship between measured variables and hidden constructs (Anderson & Gerbing, 1988). However, not all real complex problems are explained using the latent structure; specifically, if the observed items do not show any clustering characteristics. Figure 1.2 depicts the SEM and network structures.

Therefore, some problems could be efficiently represented as a network in which the most important and direct relations between the observed variables are modeled. In this dissertation, two methods will be used to model network structures from ordinal data, the *distance-based* method (Schneider et al., 1998) and *graphical lasso (glasso)* method (Tibshirani, 1996; Friedman et al., 2008). Both methods return symmetrical weight matrices. The distance-based method calculates the degree of similarity among two variables (represented by numerical vectors) and then, investigates the type of the relationship between variables (if it is direct or inverse). The graphical lasso method uses a statistical regularization technique originating in the field of machine learning to control the spurious connections in the precision matrix. In other words, this method estimates a sparser network structure, since it shrinks small partial correlation coefficients exactly to zero. Specifically, a tuning parameter λ regulates the sparsity of the network or the number of the edges. Consequently, lasso estimates a group of networks (Zhao & Yu, 2006) ranging from a fully connected network (λ_{min}) to an empty network (λ_{max}). As a result, the model that best fits into the data can be the one that minimizes the Extended Bayesian Information Criterion (EBIC; Chen & Chen, 2008). Moreover, the EBIC adds an extra penalty, the hyperparameter γ (gamma) to control the model's complexity (Foygel & Drton, 2010; Chen & Chen, 2008). To sum up, according to previous studies (Epskamp & Fried, 2017; Epskamp, 2016; Foygel & Drton, 2010), the combination of lasso regularization with EBIC model selection has been revealed the estimation of the true network structure.

1.3 The Simulation Process for Decision Making

After data modeling or estimating the network structure (learning the weights between concepts) from the available data, it is essential to perform policy scenarios through a *simulation* method which is known as *decision making* (Kosko, 1986). Consequently, decision making is a process of evaluating optimal choices under conditions of uncertainty (Muenning, 2008). Since the risks and benefits of each alternative are unknown, the decision analysis provides a mathematical framework for making decisions under conditions of uncertainly.

One of the well-known methods of modeling and simulating a system is the *Fuzzy Cognitive Map* (FCM; Kosko, 1986). FCM is classified as a soft computing technique which is trying to mimic the humans' reasoning and decision making. It combines the desirable properties of fuzzy logic and neural networks. Kosko first computed the outcome of an FCM known as FCM inference. In particular, he applied various policy cases using a computational method adapting from neural network theory (Kosko, 1986, 1992). FCMs have been applied in many scientific areas such as business, social and political sciences, ecological and environmental management, engineering, information technology, robotics, expert systems, medicine, education and so on (Papageorgiou & Salmeron, 2013; Papageorgiou & Kontogianni, 2012; Xirogiannis & Glykas, 2004; Craiger et al., 1996).

To accomplish the simulation process, first, the weights of the network are encoded into the matrix W. The initial vector V^0 at the time-step 0 ($\kappa = 0$) consists of p elements. Each element of the vector represents a value of the concept which it is induced by activating concepts of interest with value '1' or deactivating with '0' value. By applying the appropriate inference rule and the threshold function, the simulation procedure is repeatedly updating the concepts' values until i) converges in a steady-state, ii) reaches a limited cycle of steps or iii) reveals a chaotic behaviour (Papageorgiou, 2014; Papageorgiou & Salmeron, 2013; Kosko, 1986). Following the interests of the decision-maker, scenario analysis can be accomplished with either all concepts or a subset of concepts being activated depending on the problem. Therefore, different policy scenarios are determined to estimate the inference of the system (Kosko, 1986) answering the "*what—if*" conditions.

1.4 Scope and Research Objectives

To the best of our knowledge considering the interest of FCMs, none of the previous studies have introduced fully data-driven FCM models. Therefore, the general scope of this dissertation is to provide a framework for *robust data-driven models* to promote the consensus among employees. Moreover, the proposed method, the *glassoFCM with EBIC (Extended Bayesian Information Criterion) regularization* will not only assist business users to "listen to the data", understand better the domain, organize their work but also through *a* simulation process it determines valuable solutions by presenting clear insights into real-world problems cautiously and efficiently. Investigating the general goal of this dissertation, we divide it into further research objectives. Therefore, a list of specific objectives is presenting to achieve the overall aim of this dissertation. Such goals can be formalized as follows:

i) To estimate the appropriate graphical structure according to the tendency of the data to group together or not (a-priori priori analysis) and the validation of the a-priori analysis through the network analysis (post-hoc analysis).
ii) To perform fully automated data-driven models from cross-sectional ordinal data. In other words, to model network structures without the intervention of domain experts to define the relations and the weights among the nodes. Then, the estimated models derived from the glasso method and the distance-based method are compared.
iii) To evaluate the estimated models according to their accuracy and stability in order to provide robust models.
iv) To implement a free and open-source software library to run efficiently the simulation procedure.
v) To simulate the estimated model providing the ability to predict the future state of the focal node, the "Willingness of the employee to work".
vi) To identify the most important variables of the dataset which affect significantly the observed output node.

1.5 Thesis Outline

This dissertation manipulates responses from 56,246 preferences of 14,585 Belgian employees working in 15 sectors. Every participant has evaluated ten job-satisfaction features indicating their values on the scale 1 (strongly disagree) to 5 (strongly agree). Figure 1.3 depicts the focal points of each chapter while a short description of the chapters follows:

Chapter 2 describes the problem and the process of preparing the data for further investigation/analysis. Specifically, it introduces the job-satisfaction research problem, specifies the observed variables and explains the pre-processing procedure of deriving clean data. Furthermore, the demographic information of 56,246 observations according to gender, education, activity, situation and working sector is presented (Dikopoulou et al., 2020b).

Chapter 3 introduces the Fuzzy Cognitive Map (FCM) methodology focusing on the simulation process (Kosko, 1986, 1992). First, the literature review on previous studies in the business domain is stated. Then, a detailed description of our developed R package, the '*fcm*' is presented (Dikopoulou & Papageorgiou, 2017; Dikopoulou et al., 2018). It is available in CRAN (Comprehensive R Archive Network; *https://cran.r-project.org/web/packages/fcm/*) with a variety of adjustments. Particularly, the '*fcm*' supports six inference rules, six threshold functions and a range of values for parameters λ and ε. Particularly, λ determines the steepness of the transformation function and ε the residual error in every iteration of the FCM inference. Furthermore, this chapter points out the analytical step by step examples including the correspondent code, the input arguments and the output results of the '*fcm.infer*' function (of the 'fcm' Package). Last, the recommendations to the users are provided to assist them about the use of the 'fcm' package.

Chapter 4 describes the definition of a network and its classifications. Then, nine algorithms are applied to the data to investigate the underlying latent variables from the set of observed variables or not (a priori analysis). The results show that the variables do not present clustering preferences and can be modelled as graphs in which each variable is not expressed by other unobserved measures (factors). For this reason, two methods that model data into graph structures are introduced, the distance-based method (Schneider et al., 1998) and the graphical least absolute shrinkage and selection operator (glasso) with the Extended Bayesian Information Criterion (EBIC; Chen & Chen, 2008), or simply glasso with EBIC regularization (Friedman et al., 2008).

The distance-based method has been used to create FCM structures based on numerical measurements. In brief, the variables are represented as numerical vectors which are transformed into fuzzy vectors. Next, the determination of positive or negative relations among concepts are defined. Afterwards, the strength (the weight) of each edge is calculated based on the distances between numerical vectors. At last, experts must decide the direction of causality between concepts indicating the final FCM weighted matrix. Consequently, this method is categorized as a semi-automated method since limited human intervention is required. On the other

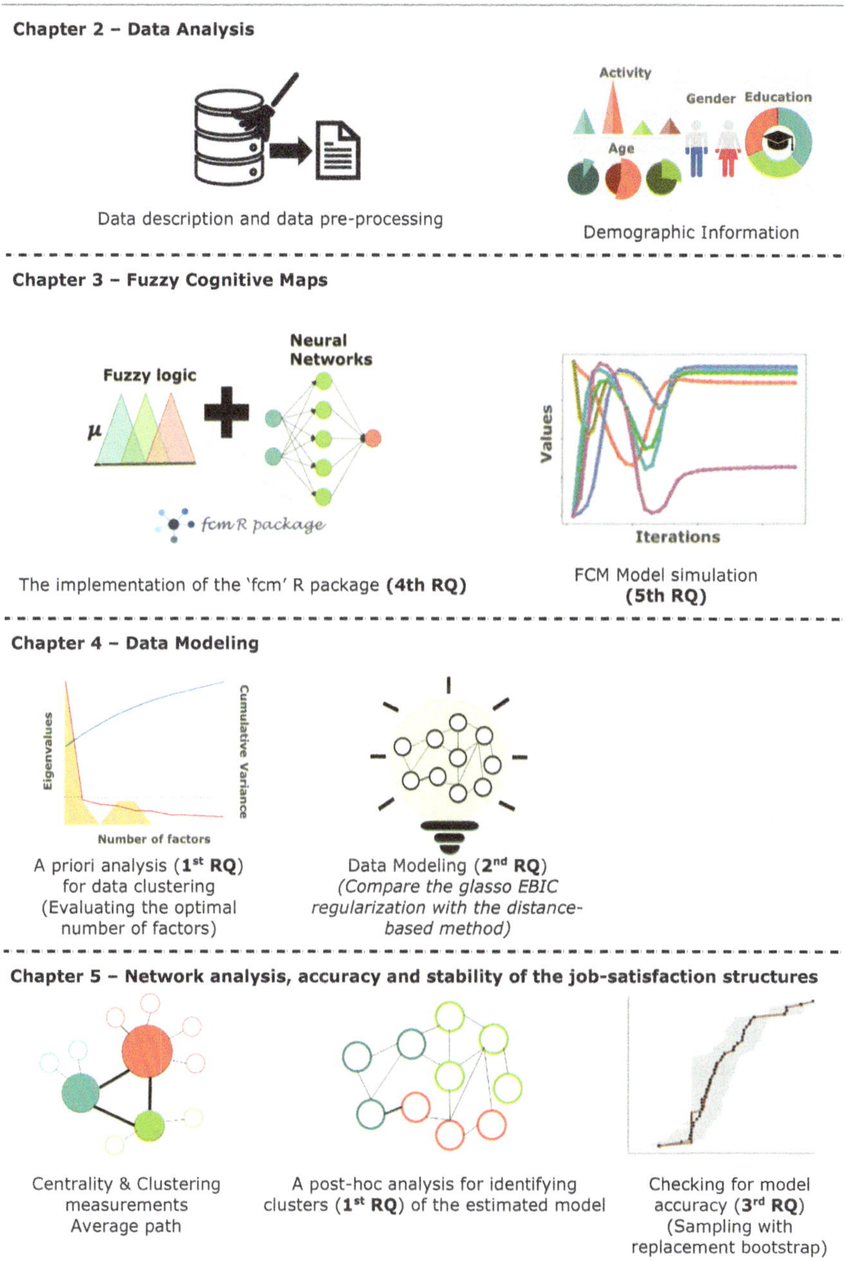

Fig. 1.3 The highlights of each chapter. **RQ** denotes the *Research Question* as described in *Scope and Research Objectives*

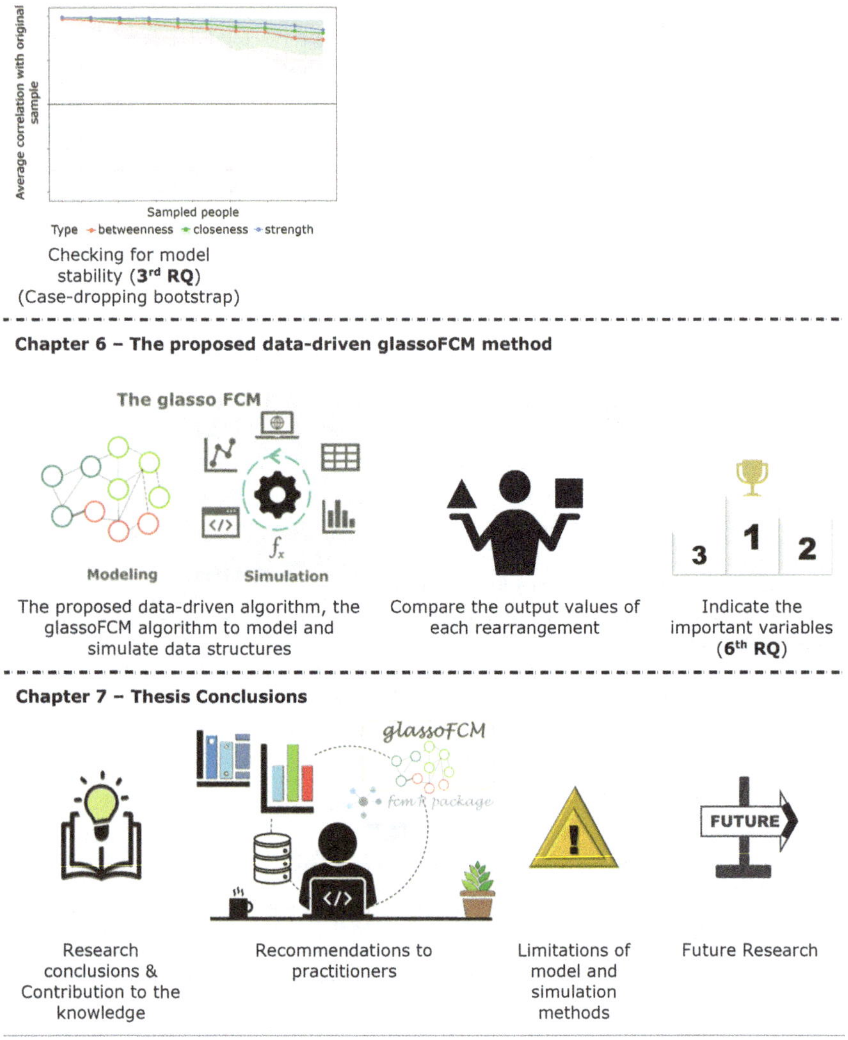

Fig. 1.3 (continued)

hand, regularized glasso is an algorithm which describes the dependencies between a set of variables. It belongs to a family of probabilistic graphical models and it is used to learn the structure of a probabilistic graphical model from observational data to elucidate the relationships among variables. The glasso method utilizes the ℓ_1 regularization parameter to control the sparsity of the precision matrix. Consequently, it estimates a group of networks; specifically, 100 models (Zhao & Yu, 2006) ranging from a fully connected network (λ_{min}) to an empty network (λ_{max}). Therefore, the best network out of this range of networks must be selected by optimizing the fit of the network to the data. According to previous studies (Foygel

& Drton, 2011; Epskamp, 2016; Epskamp & Fried, 2017), minimizing the EBIC, the true network structure was estimated indicating that high specificity has occurred. Moreover, the EBIC added an extra penalty, the hyperparameter γ (gamma) to control (additionally) the sparsity of the model (Zhao & Yu, 2006; Foygel & Drton, 2011).

The derived models from both methods are compared (Dikopoulou et al., 2017) according to their density (the total number of edges), statistical tests (examining the significant difference) and time performance of the algorithms (in seconds). Results show that glasso method estimates faster and sparser graphs with a higher standard deviation on edge weights comparing to a distance-based method. For all these reasons, the glassoFCM models are efficient and easier to interpret.

Chapter 5 overviews network analysis approaches to determine and interpret meaningful measures of local and global network indices. Specifically, centrality indices (strength, closeness and betweenness centrality) are used to identify the most important nodes in the network (Dikopoulou et al., 2020b), clustering coefficients are applied to distinguish redundant nodes and the small-worldness (SW) property is examined to determine if the observed nodes present high clustering features and form or not separate groups (post-hoc analysis). Finally, 22,500 nonparametric bootstraps (sampling with replacement) are applied to assess the accuracy of the business network parameters. At last, 22,500 case-dropping bootstraps are performed to test the stability of the estimated networks.

Since most of the studies have been focused on semi-automated and automated methods to learn FCMs based on historical data, few approaches have been applied to define the relations and the weights among concepts (nodes) from cross-sectional data (not historical data). ***Chapter 6*** presents the proposed data-driven *glassoFCM* algorithm with EBIC regularization (Dikopoulou et al., 2017) to model and simulate cross-sectional datasets. The glassoFCM method is a combination of two well-known methods, the *regularized graphical lasso* which models the data (Chapter 4) and the *FCM* which simulates decisions and/or predictions (Chapter 3). This chapter demonstrates how an undirected weighted matrix is transformed into a directed weighted matrix in order to making-decisions for a single output variable (using the simulation process of the '*fcm.infer*' function; Dikopoulou & Papageorgiou, 2017). Different rearrangements of the input variables according to six cases/rankings (i) strength, ii) closeness, iii) betweenness, iv) average of variables in the datasets, v) random and vi) the inverse of the strength) and several scenarios are performed to investigate if the value of the outcome variable after the FCM inference process changes significantly. In total, 1170 scenarios are simulated to indicate not only the deviation of the output values but also to ascertain the most important factors that influence the output concept ('willingness to work') of the FCM (Dikopoulou et al., 2020a, b).

Chapter 7 summarizes the highlights of this dissertation. Specifically, different aspects will be discussed, including the research conclusions, the contribution to knowledge, the recommendations to experts, the limitations of this dissertation and finally, the future research.

References

Anderson, J. C., & Gerbing, D. W. (1988). Structural equation modelling in practice: A review and recommended two-step approach. *Psychological Bulletin, 103*(3), 411–423.

Brunton, B. W., & Beyeler, M. (2019). Data-driven models in human neuroscience and neuroengineering. *Current Opinion in Neurobiology, 58*, 21–29. https://doi.org/10.1016/j.conb.2019.06.008

Chen, J., & Chen, Z. (2008). Extended bayesian information criteria for model selection with large model spaces. *Biometrika, 95*(3), 759–771.

Craiger, J. P., Goodman, D. F., Weiss, R. J., & Butler, A. (1996). Modeling organizational behavior with Fuzzy Cognitive Maps. *International Journal of Computational Intelligence and Organisations, 1*, 120–123.

Dikopoulou, Z., & Papageorgiou, E. (2017). *Inference of Fuzzy Cognitive Maps (FCMs)*. The Comprehensive R Archive Network (CRAN), R package Version 0.1.3. https://cran.r-project.org/web/packages/fcm/vignettes/vignettes.html

Dikopoulou, Z., Papageorgiou, E., Jetter, A., & Bochtis, D. (2018). *Open-source tool in R language to estimate the inference of the Fuzzy Cognitive Map in environmental decision making*. 9th International Congress on Environmental Modeling and Software, Fort Collins, USA, June 24–28, 2018.

Dikopoulou, Z., Papageorgiou, E., Mago, V., & Vanhoof, K. (2017). A new approach using mixed graphical model for automatic design of fuzzy cognitive maps from ordinal data. *IEEE International Conference on Fuzzy Systems (FUZZ-IEEE)*. https://doi.org/10.1109/fuzz-ieee.2017.8015607

Dikopoulou, Z., Papageorgiou, E., & Vanhoof, K. (2020a). *Retrieving sparser fuzzy cognitive maps directly from categorical ordinal dataset using graphical lasso models and the max-threshold algorithm*. IEEE International Conference on Fuzzy Systems (FUZZ-IEEE).

Dikopoulou, Z., Papageorgiou, E., & Vanhoof, K. (2020b). *From undirected structures to directed graphical lasso fuzzy cognitive maps using ranking-based approaches*. IEEE International Conference on Fuzzy Systems (FUZZ-IEEE).

Epskamp, S., & Fried, E. I. (2017). A tutorial on regularized partial correlation networks. *Psychological Methods* (In press). https://doi.org/10.1037/met0000167.

Epskamp, S. (2016). *Regularized Gaussian psychological networks: Brief report on the performance of extended BIC model selection*. arXiv preprint, arXiv:1606.05771.

Foygel, R., & Drton, M. (2010). Extended Bayesian information criteria for Gaussian graphical models. *Advances in Neural Information Processing Systems, 23*, 2020–2028.

Foygel, R., & Drton, M. (2011). Bayesian model choice and information criteria in sparse generalized linear models. arXiv preprint, arXiv:1112.5635.204.

Friedman, J. H., Hastie, T., & Tibshirani, R. (2008). Sparse inverse covariance estimation with the graphical lasso. *Biostatistics, 9*(3), 432–441.

Hoppock, R. (1937). Job satisfaction of psychologists. *Journal of Applied Psychology, 21*(3), 300–303. https://doi.org/10.1037/h0057579

Kosko, B. (1986). Fuzzy cognitive maps. *International Journal on Man-Machine Studies, 24*, 65–75.

Kosko, B. (1992). *Neural networks and fuzzy systems: A dynamical systems approach to machine intelligence*. Prentice-Hall.

Locke, E. A. (1970). Job satisfaction and job performance: A theoretical analysis. *Organizational Behavior and Human Performance, 5*(5), 484–500. https://doi.org/10.1016/0030-5073(70)90036-x

Muenning, P. (2008). Decision analytic modeling. *International Encyclopedia of Public Health*, 71–76. https://doi.org/10.1016/b978-012373960-5.00332-4.

Newman, M. E. J. (2010). *Networks: An introduction* (pp. 3, 5, 6, 34, 77, 195, 205, 208). Oxford University Press.

Papageorgiou, E. I., & Kontogianni, A. (2012). Using fuzzy cognitive mapping in environmental decision making and management: A methodological primer and an application. In S. S. Young & S. E. Silvern (Eds.), *International perspectives on global environmental change* (pp. 427–450). InTech.

Papageorgiou, E. I., & Salmeron, J. L. (2013). A review of fuzzy cognitive maps research during the last decade. *Fuzzy Systems, IEEE Transactions, 21*, 66–79.

Papageorgiou, E. I. (2014). Fuzzy cognitive maps for applied sciences and engineering from fundamentals to extensions and learning algorithms. *Intelligent Systems Reference Library, 54*.

Schneider, M., Shnaider, E., Kandel, A., & Chew, G. (1998). Automatic construction of FCMs. *Fuzzy Sets and Systems, 93*(2), 161–172.

Spector, P. E. (1997a). Advanced topics in organizational behavior. In *Job satisfaction: Application, assessment, causes, and consequences*. SAGE. https://doi.org/10.4135/9781452231549

Spector, P. E. (1997b). *Job satisfaction: Application, assessment, causes and consequences*. SAGE.

Tang, S., Siu, O., & Cheung, F. (2014). A study of work–family enrichment among Chinese employees: The mediating role between work support and job satisfaction. *Applied Psychology, 63*(1), 130–115.

Thompson, B. (2007). Factor analysis. *The Blackwell Encyclopedia of Sociology*. https://doi.org/10.1002/9781405165518.wbeosf003

Thurstone, L. L. (1931). Multiple factor analysis. *Psychological Review, 38*(5), 406–427.

Tibshirani, R. (1996). Regression shrinkage and selection via the lasso. *Journal of the Royal Statistical Society. Series B (Methodological), 58*, 267–288.

Weiss, H. M., & Merlo, K. L. (2015). Job satisfaction. *International Encyclopedia of the Social & Behavioral Sciences*, 833–838. https://doi.org/10.1016/b978-0-08-097086-8.22029-1

Xirogiannis, G., & Glykas, M. (2004). Fuzzy cognitive maps in business analysis and performance-driven change. *IEEE Transactions on Engineering Management, 51*(3), 334–351. https://doi.org/10.1109/tem.2004.830861

Zhao, P., & Yu, B. (2006). On model selection consistency of lasso. *The Journal of Machine Learning Research, 7*, 2541–2563.

Chapter 2
Data Analysis

Nowadays, companies in the business world are struggling to be more profitable and effective, reacting faster and offering a higher quality of products and services by minimizing every reason that induces the company's cost. Moreover, the available information for decision making has exploded in the past few years, in which some of these datasets are analyzed with different statistical techniques, while others with complex, multivariate statistical methods such as machine learning techniques to convert messy data into useful knowledge. In this chapter, the job satisfaction problem will be described and will be visualized. Specifically, the job-satisfaction variables and the process of collecting the evaluations from employees will be explained. In total, 94,135 observations "follow" the cleaning process to derive reliable and consistent data and then, the most significant data insights will be visualized. Particularly, demographic visualizations of the clean dataset will be displayed to inform the audience about the collection and analysis of the general characteristics of specific groups of people such as gender, age, activity, education and working sectors. Next, ordinal data in a five-point Likert-type scale is plotted to produce interesting summaries about the employees' responses in order to classify which of the observed variables is the most or least attractive. Then, the associations among the variables are displayed as networks in order to determine fast interesting patterns, which are hard to identify directly from the data. Finally, we attempt to infer the most significant pairs of variables in each working sector that affect the job-satisfaction problem.

2.1 Introduction

Many researchers have been studying the attractiveness of a company in a business domain, but this issue will continue to be under investigation since the employees' demands are changing decade after decade (Weiss and Merlo 2015; Tang et al., 2014; Latham and Budworth, 2007; Spector 1997; Hoppock, 1935). The term

Z. Dikopoulou, *Modeling and Simulating Complex Business Perceptions*, Fuzzy Management Methods, https://doi.org/10.1007/978-3-030-81496-0_2

attractiveness refers to the case where a company focuses only on employee' preference to boost the employee's satisfaction. As a consequence, the company retains the personnel and attracts new efficient employees (Locke, 1970). The sales or services will be increased and finally; the company remains competitive in the market. These effects can be achieved if the company improves the observed variables that participants considered more important. In this chapter, human reasoning extraction is investigated to evaluate the attractiveness of each company resident in Belgium. Furthermore, the goals of this chapter are: 1) to introduce the variables of the research problem, 2) to obtain reliable and consistent data, 3) to visualize the aspects of the data for higher interpretation and 4) to show the associations among the observed variables.

2.2 Data Description

During the data acquisition, 14,585 Belgian respondents evaluate ten questions (features, attributes or observed variables) related to the job satisfaction problem indicating their values on the scale of 1 (strongly disagree) to 5 (strongly agree). Table 2.1, presents the set of the observed variables of the dataset as it was structured in the questionnaire. Moreover, the questionnaire includes demographic questions that each participant has to complete such as *gender* (male, female), *age* (18–24, 25–44, 45–65), *education* (primary/secondary, bachelor, master), *activity* (production worker, office worker, manager) and *working sector* (Automotive (S_1), Constructor (S_2), Finance (S_3), Food (S_4), Human Resources (S_5), Manufacturing (S_6), Informatics (S_7), Non-profit (S_8), Pharmaceutical/Chemical Industry (S_9), Energy Services (S_{10}), Retail (S_{11}), Business Services (S_{12}), Transportation/Logistics/ Wholesales (S_{13}), Travel/Leisure/Hospitality (S_{14}), Other (S_{15})). The questionnaire was managed by marketing experts from Randstad (*http://www.randstad.com*), the second-largest HR provider in the world. It expanded its operations to 39 countries, representing more than 90% of the global HR services market.

After completing the demographic questions, a list of 349 Belgian companies is distributed to each participant requesting to indicate which of these 349 companies he or she is aware of. Then, according to the responses, each correspondent evaluates the job-satisfaction variables (V_1–V_{11}) as presented in Table 2.1 for the selected companies from strongly disagree to strongly agree. Figure 2.1 demonstrates the required steps (step 1–step 4) that are necessary to complete the evaluation of the job-satisfaction variables. First, an employee from the constructor sector (step 1) indicates the number of known companies (step 2). As it is observed, he selects only two companies. Then, he assesses the features in a five-point Likert scale for each company (step 3); therefore, he will evaluate 20 job satisfaction variables (10 for each company). In total, 94,135 observations are collected from 8921 participants (step 4); consequently, an ordinal cross-sectional dataset is formed.

Ordinal data is a classification of categorical data and it incorporates the rank-ordering operation (Stevens, 1946). In addition, ordinal data are used widely by

Table 2.1 The employee satisfaction questionnaire is formatted in a five-point Likert scale. The first ten variables (V_1–V_{10}) reflect the input features correlated to the job-satisfaction problem and the last variable (V_{11}) indicates the output (the decision) variable

Employee Satisfaction Survey

Considering the company that you had selected,

"How important are the factors (V_1 - V_{10}) in order to work to this company ?"

"What is your willingness to work to that company (V_{11}) ?"

Please rate the following statements:

		Strongly Disagree	Disagree	Neither Agree nor Disagree	Agree	Strongly Agree
V_1	Competitive salary package	○	○	○	○	○
V_2	Prospects/career opportunities	○	○	○	○	○
V_3	Pleasant working environment	○	○	○	○	○
V_4	Offers long-term job security	○	○	○	○	○
V_5	Good balance between private life & work	○	○	○	○	○
V_6	Interesting jobs	○	○	○	○	○
V_7	Financially sounds	○	○	○	○	○
V_8	Good quality of training	○	○	○	○	○
V_9	Strong management	○	○	○	○	○
V_{10}	Deliberately handles the environment and society	○	○	○	○	○
V_{11}	Willingness to work to the company	○	○	○	○	○

researchers to study judgments, feelings or emotions of people. A classic example of an ordinal scale is the Likert scale. A Likert scale is usually a five to ten-point scale with different opinions that fluctuate from one extreme to the other, such as Very Bad, Bad, Mediocre, Good, Very Good. Consequently, we model a glassoFCM graph according to employees' perspectives using a five-point Likert scale on ten observed variables.

2.3 Data Pre-processing

After placing data into rows and columns in a table structure to form a cross-sectional ordinal dataset, a time-consuming but important step before initiate any analysis procedure is called *data examination* or *data cleaning* (Hair et al., 2014; Salkind, 2010; Fayyad et al., 1996). Moreover, in statistics, the cleaning procedure is a crucial process, since it is impossible to accomplish an accurate statistical analysis of inaccurate data. This is known as "*Garbage In Garbage Out* (GIGO)" (The Hammond Time, 1957). Therefore, dirty or coarse data are deleting or in many

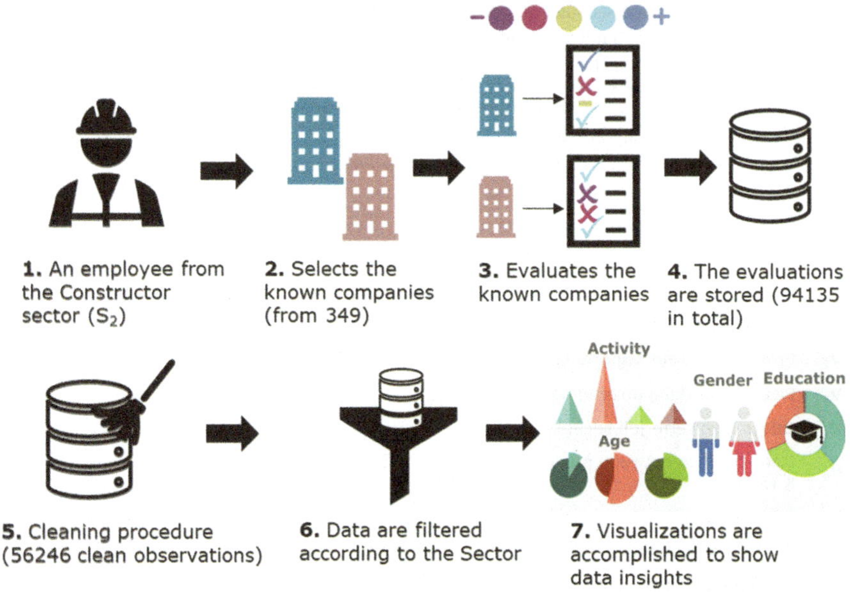

Fig. 2.1 The procedure of assessment of the job-satisfaction variables for specific companies. Then, follows the cleaning procedure, the filtering process and finally, the required visualizations

cases replaced or modified to prevent fundamental errors in further analysis techniques. Missing data and outliers are identified in order to obtain clean and consistent data.

In our dataset, extraordinary observations (outliers) are not detected; nevertheless, unengaged respondents are noticed. Generally, the unengaged respondents can appear in two cases when the participants a) enter the same value in their responses; consequently, the standard deviation equals to 0 and b) use different patterns to complete the questionnaire, such as '1, 2, 3, 4, 5' (Fayyad et al., 1996). From the cleaning procedure (step 5 of Fig. 2.1), 37,889 observations were identified as unengaged and were removed from the dataset. Therefore, the dataset contains 56,246 valid surveys (a response rate of 59.75%) which will drive to significant data-driven insights.

2.4 Data Visualizations

From this survey, the preference of participants from different gender, age, education, activity and working sectors were evaluated to analyze the employees' reasoning, when they were applying for a job position. Figure 2.2 demonstrates the demographic information of the whole sample, which 30,185 are male and 26,061

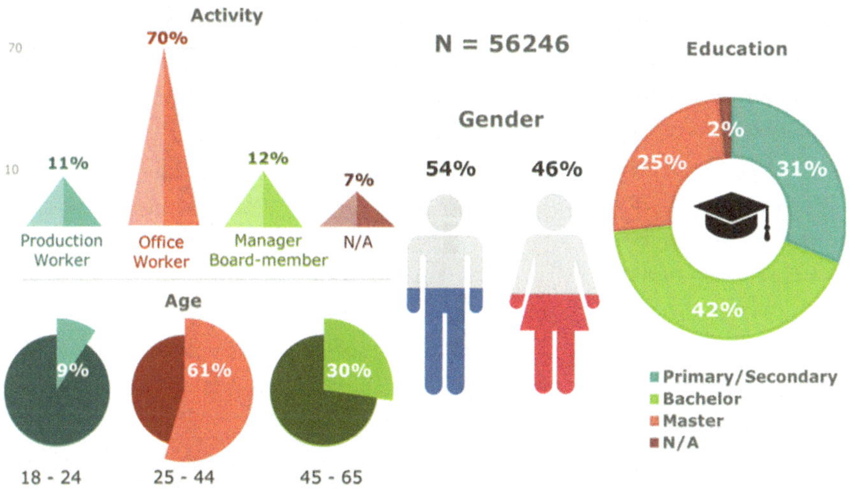

Fig. 2.2 The demographic information (gender, age, activity and education) of the clean dataset, including 54,246 employees who work in 15 working sectors in Belgium

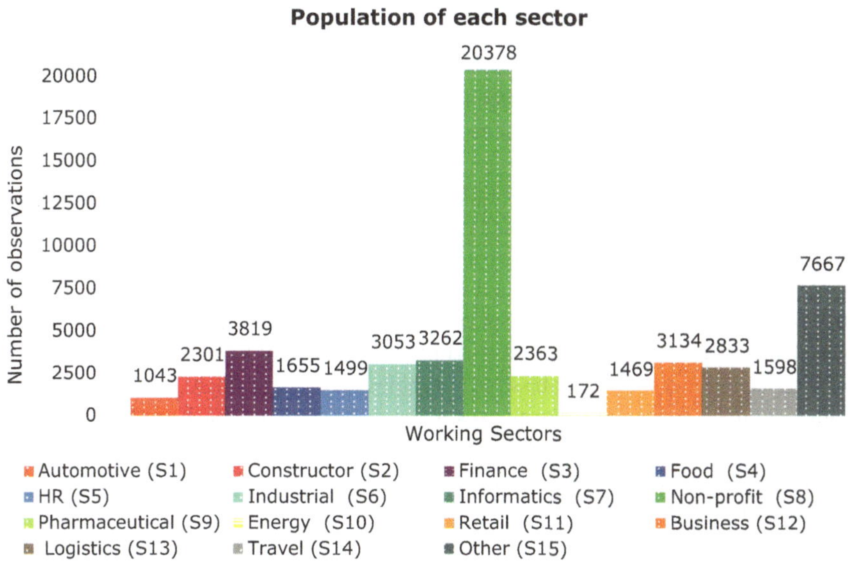

Fig. 2.3 The number of observations of the 15 working sectors ($S_1 - S_{15}$) in Belgium

are female employees. Furthermore, the majority of the correspondents are between 25 and 44 years old (61%), bachelor holders (42%) and office workers (70%).

Figure 2.3 depicts a population of 15 working sectors. As it is observed, the Non-profit (S_8) sector depicts the higher percentage (36.2%) of the total observations while the sector, Energy services (S_{10}), contains the small sample (0.3%) of the

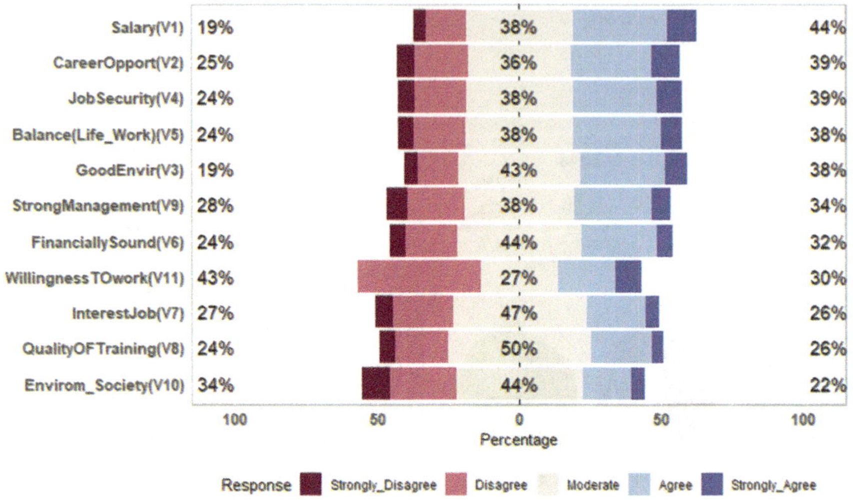

Fig. 2.4 A total of 56,246 respondents replied to the survey concerning job satisfaction. Each employee evaluated one of five levels of agreement or disagreement of 11 variables

records. Particularly, most of the sectors include less than 4000 observations with the exclusion of sectors Others (S_{15}) and Non-profit S_8. Furthermore, the demographic information for each working sector is presented in the Table A.1 of the Appendix A. Even if Fig. 2.2 depicts a balance between males and females, in Table A.1 are detected the sectors that obtain more males comparing to females which are: Automotive (70.7%), Constructor (73.2%), Industry (71.2%), Informatics (78.7%), Pharmaceutical (61.0%) and Logistics (67.4%). On the contrary, more females are inspected in HR (68.3%) and Energy services (65.1%).

Moreover, in order to analyze and visualize the Likert response items, we applied the 'likert' R package (Bryer and Speerschneider, 2016). Thus, the stacked bar plot is selected to present the Likert results (Figs. 2.4 and 2.5). Figure 2.4 illustrates the ordinal preferences of the observed variables (V_1–V_{11}). Each panel of the plot indicates the percentage of responses to each level of satisfaction. For example, dark-blue color denotes the ordered categorical scale "Strongly Agree"; while the dark purple color determines the scale "Strongly Disagree". Furthermore, the consensus of negative responses ("Strongly Disagree" and "Disagree"), neutral ("Moderate") and positive responses ("Strongly Agree" and "Agree") are summarized and presented. For instance, the first variable, Competitive Salary (V_1) in Fig. 2.4 includes 19% pessimistic responses, 38% neutral and 44% optimistic responses. Alternatively, 44% of the positive responses indicate that 31,497 participants ascertain that the salary is a preferable feature when they are looking for a job. Additionally, it is inferred that except V_1, Prospects/career opportunities (V_2) and Offers long-term job security (V_4) concentrate the most favorable responses in the total survey (Fig. 2.4). Therefore, they are ordered according to the percentage of positive responses.

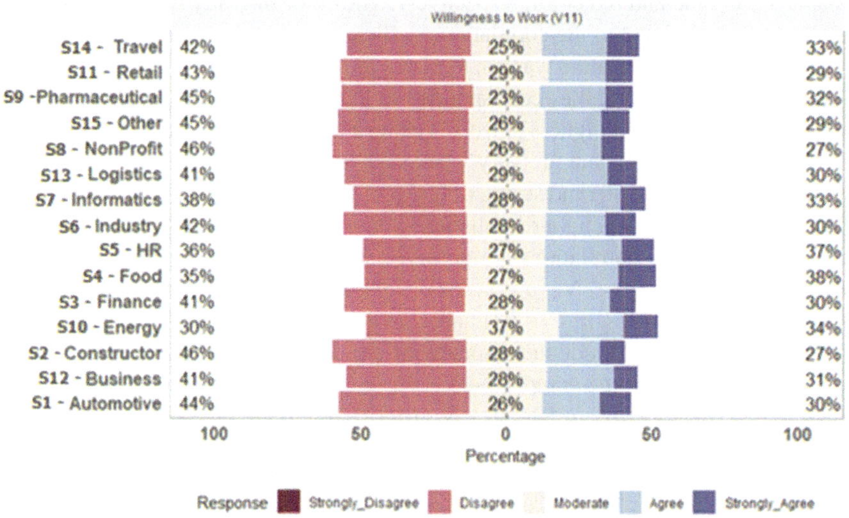

Fig. 2.5 The Likert-type scales of the ordinal responses in the feature V_{11} (Willingness to work) in every sector. It is observed that no one of the employees responds to this question with the "Strongly Disagree" scale; nevertheless, the negative effect is visible

On the contrary, the *Willingness of the employee to Work* (V_{11}) is rated negatively (43%) comparing to the neutral (27%) and the positive (30%) groups (Fig. 2.4). In fact, V_{11} is not a feature to measure the employee satisfaction following the three main relationships between employee and organization, supervisor and co-worker proposed by Tang et al. (2014). Typically, V_{11} reflects the personal evaluation about the willingness of the employee to work which it is depended from the job-satisfaction variables (V_1–V_{10}). For this reason, the preferences of V_{11} are investigated separately in the observed sectors (Fig. 2.5). As it is depicted in Fig. 2.5, none of the employees evaluated the V_{11} with strongly negative scale (since the "Strongly Disagree" scale is 0). Nevertheless, the "*Disagree*" measurement is the most selected choice of the majority. The only exceptions are observed in three sectors: HR (S_5), Food (S_4) and Energy services (S_{10}), in which the differences between positive and negative responses are not significant. We assume that the negative impact of V_{11} stems from the nature of humans to work to retain their livelihood at a good level and not to maintain a good standard of living in order to work.

The linearity of the ordinal variables is examined with the use of a nonparametric measure named as Spearman Bivariate Association is applied to the ordinal datasets (Spearman, 1904). The Spearman correlations are performed with '*phych*' R package (Revelle, 2018). For the graphical displays of the Spearman correlation matrices, the visualization method '*corrplot*' was applied using the homonym '*corrplot*' package in R (Wei et al., 2017). Thus, positive and negative correlations are displayed in blue and red color, respectively. The color intensity and the size of the circle are proportional to the correlation coefficients.

A Spearman's correlation was run to assess the relationship between the observed variables (ten input variables corresponded to the job satisfaction problem and one focal variable, the "Willingness to work") using a large sample of 56,246 employees. The results show that there are very weak ($.00 < \rho < .20$), weak ($.20 \leq \rho < .40$), moderate ($.40 \leq \rho < .60$) and strong ($.60 \leq \rho < .80$) positive correlations among the observed variables, which were statistically significant, $.26 \leq \rho \leq .64$, $p = .000$. Specifically, strong correlations are observed between *Interesting jobs* with *Good quality of training* ($V_7 - V_8$, $\rho_{78} = .64$), *Competitive salary package* and *Career opportunities* ($V_1 - V_2$, $\rho_{12} = .63$) and *Career opportunities* and *Interesting jobs* ($V_2 - V_7$, $\rho_{27} = .61$). Moderate associations are determined among *Career opportunities* and *Quality of training* ($V_2 - V_8$, $\rho_{28} = .59$), $V_6 - V_9$ ($\rho_{69} = .55$), $V_3 - V_7$, ($\rho_{37} = .54$), $V_2 - V_4$ ($\rho_{24} = .54$), $V_1 - V_2$, ($\rho_{17} = .54$), $V_1 - V_8$, ($\rho_{18} = .53$), $V_4 - V_6$ ($\rho_{46} = .54$).

Conversely, the weakest associations ($\rho \leq .30$) are observed mainly among *Competitive salary package* and *Deliberately handles the environment and society* ($V_1 - V_{10}$, $\rho_{1,\,10} = .28$) and *Willingness to work* with the variables *Good balance between private life & work* ($V_{11} - V_5$, $\rho_{11,\,5} = .28$), *Financially sound* ($V_{11} - V_6$, $\rho_{11,\,6} = .28$), *Strong management* ($V_{11} - V_9$, $\rho_{11,\,9} = .26$), and *Deliberately handles the environment and society* ($V_{11} - V_{10}$, $\rho_{11,\,10} = .30$). On the whole, the variables with higher average values are the *Prospects/career opportunities* (V_2; $\mu_2 = .49$, *st. dev* = .101) and *Interesting jobs* (V_7; $\mu_7 = .49$, *st. dev* = .082).

Generally, a correlation matrix that describes pairwise associations can be visualized as a network in which each variable is a node (vertex) and each correlation an edge (link). In addition, each edge carries a specific weight which in this case is the association strength among two variables. Moreover, stronger associations are represented by wider and more saturated edges; while blue and red links signify positive and negative correlations, respectively. In fact, correlation networks are helpful for a quick browse through the data by highlighting interesting patterns in which are hard to identify (Costantini et al., 2015). Therefore, Fig. 2.7 depicts the network structure from the correlogram (association matrix) that is presented in Fig. 2.6. For this network visualization, the R package 'qgraph' R package is applied (Epskamp et al., 2012). More information about networks is provided in Chaps. 3 and 4.

However, the weighted network in Fig. 2.7 is fully connected (every node is connected with all the other nodes). This outcome strengthens our suspicion that the associations among nodes may be spurious increasing the number of false-positive connections. To recap, two nodes are linked if the weight of the edge between them is nonzero and two nodes are not connected is the association weight value is zero. According to Costantini et al. (2015), association networks are usually fully connected networks since they are based on an empirical sample and return arbitrarily small weights between variables. Furthermore, Appendix B presents the Spearman's correlations (Fig. B.1) of each sector which the returned matrices show fully connected structures. Thus, these associations among the pair of variables

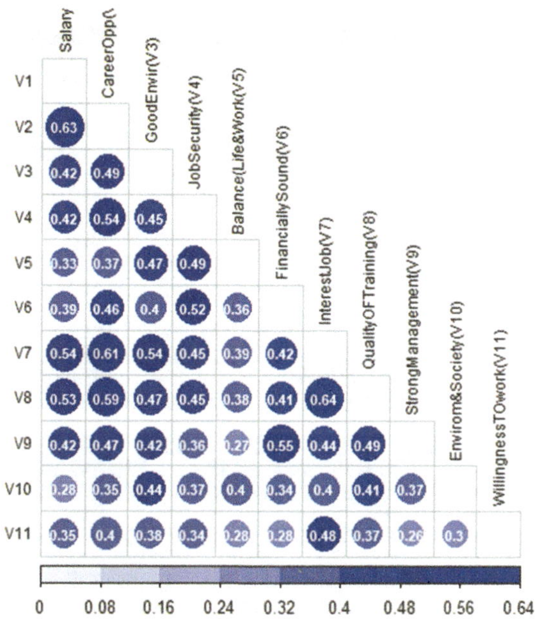

Fig. 2.6 The visualization of the correlation matrix of 56,246 observations to show the associations among 11 variables. Blue and saturated circles indicate positive and stronger associations

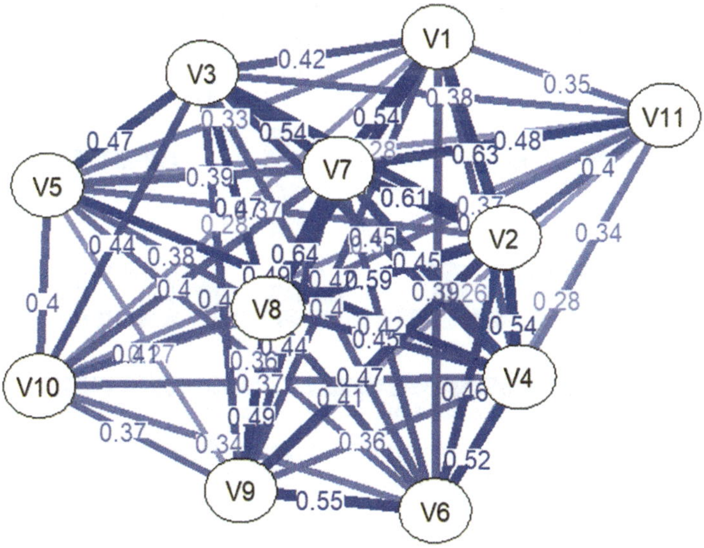

Fig. 2.7 Low and medium positive bivariate correlations ($0.26 \leq r_s \leq 0.64$) are identified from the dataset of 56,246 observations. The blue color of the lines indicates positive associations; while, the strength of the connections is visualized with thick lines for strong associations and with less saturated lines for weaker connections

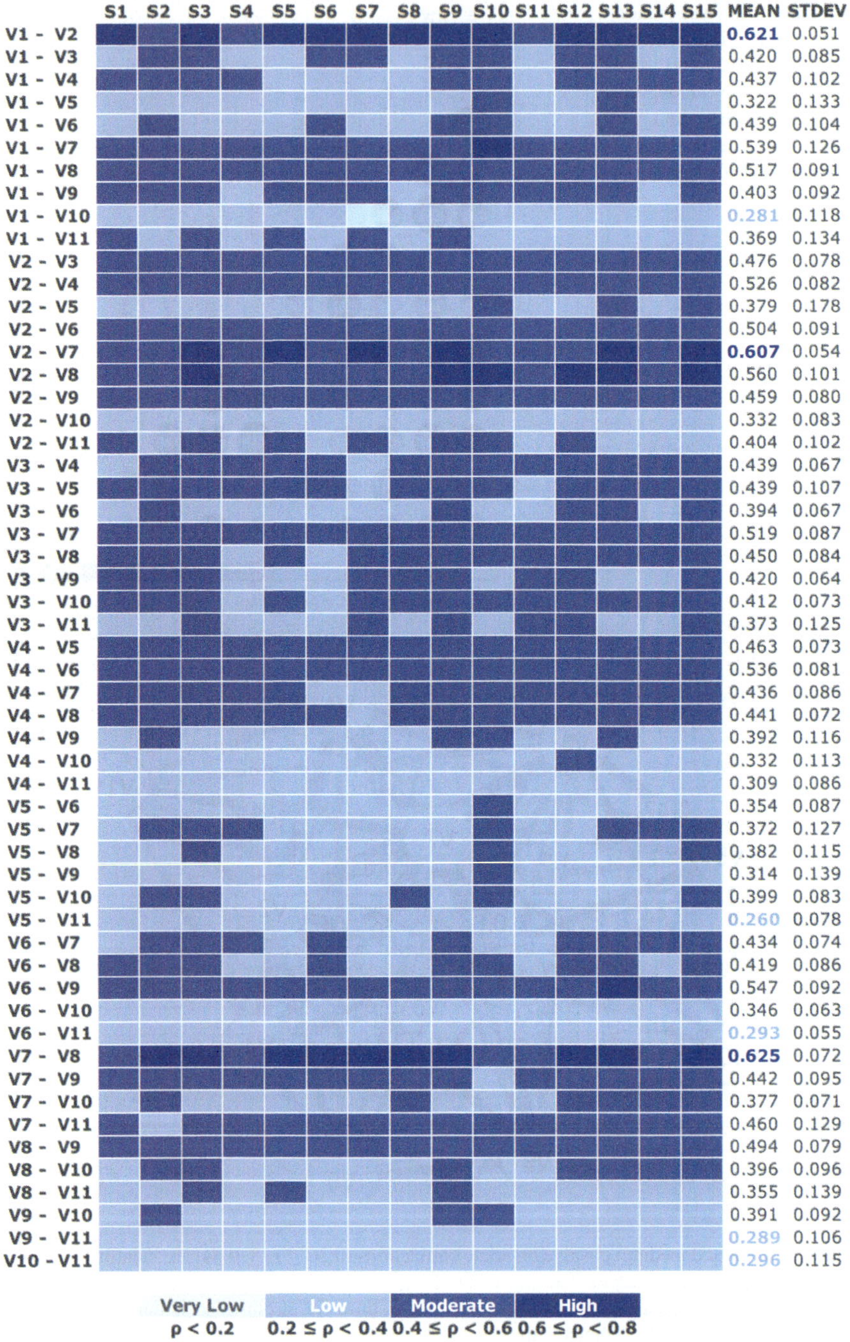

Fig. 2.8 Plot visualization of Spearman associations between two variables in every sector. Rows represent the pairs of variables and columns depict the working sectors

are summarized in Fig. 2.8. Every weight of the plot is colored according to its strength. Specifically, there are four coloring types: very low ($0 \leq \rho < 0.2$), low ($0.2 \leq \rho < 0.4$), moderate ($0.4 \leq \rho < 0.6$) and high ($0.6 \leq \rho < 0.8$). Weaker correlations are displayed with a light blue color; while, higher correlations use darker blue pigment.

In total, the pairs of variables with the higher connections (<0.6) are observed between *Interesting jobs* and *Quality of Training* ($V_7 - V_8$; 0.625 ± 0.072), *Competitive salary package* and *Prospects/career opportunities* ($V_1 - V_2$; 0.621 ± 0.051) and *Career opportunities* and *Interesting jobs* ($V_2 - V_7$; 0.607 ± 0.054). In contrast, the weakest connections (<0.3) are noticed among *Competitive salary package* and *Dealing with the environment and society* ($V_1 - V_{10}$; 0.260 ± 0.078) and *Willingness to work* with Good balance between private life ($V_{11} - V_5$; 0.260 ± 0.078), Strong management ($V_{11} - V_9$; 0.289 ± 0.106), Financially sound ($V_{11} - V_6$; 0.293 ± 0.055) and *Dealing with the environment and society* ($V_{11} - V_{10}$; 0.296 ± 0.115). Moreover, the lowest association is detected between $V_1 - V_{10}$ ($\rho_{1,10}^{S7} = 0.197$) in Informatics (S_7); while the highest associations are observed among $V_7 - V_8$ in Finance (S_3; $\rho_{7,8}^{S3} = 0.677$) and Pharmaceutical (S_9; $\rho_{7,8}^{S9} = 0.673$).

Nevertheless, the condition of the fully-connected graph affects negatively the interpretability of the network making the interpretation of the structure more difficult (Scott, 2000). Contrary to our expectations, we did not investigate interpretable models to explain the data insights due to very high connectivity among the variables. Therefore, a more efficient method is required to estimate an explainable network minimizing the spurious connections to zero.

Next chapter will introduce a technique originate from machine learning to solve this problem.

2.5 Conclusions

This chapter evaluated the preference of participants to analyze the employees' reasoning when they were applying for a job position. Particularly, ten variables were measured in a five-point Likert scale. The data cleaning was an essential procedure to improve the data quality and therefore, more accurate employee information was collected. Since it is known that a picture is worth a thousand words, demographic information concerning different groups such as gender, age, education, activity and working sector were visualized to obtain better data insights. The main goal of this chapter was to "listen" and investigate what employees considered as important factors that need to be improved to increase of keep the personnel satisfied. For this reason, we attempted to model the employees' evaluations in 15 association networks that represented the datasets of 15 working sectors. Nevertheless, correlation networks may not be optimal since spurious associations among nodes can be estimated in which true relations between nodes are not revealed. As a result, the fully-connected structures affect negatively the interpretability of the networks since they do not provide clear insights. Consequently, a more optimized method is required to estimate interpretable networks minimizing the spurious connections to zero producing sparser and accurate networks.

The next chapter presents a well-known method for modeling and simulating complex systems, the Fuzzy Cognitive Maps (FCMs). In this dissertation, the FCM methodology will be used only for the simulation process. Chapter 4, introduces two methodologies to model cross-sectional datasets. The first one is a technique that originated from the machine learning which is known as glasso with EBIC regularization and the other is the distance-based method. Specifically, these methods have been used in previous studies to model data structures. For the best of our knowledge, we first compare these two methodologies, the glasso and the distance-based method to indicate which of them is the most efficient algorithm.

References

Bryer, J., & Speerschneider, K. (2016). *likert: Analysis and visualization Likert items*. https://cran.r-project.org/web/packages/likert/index.html

Costantini, G., Epskamp, S., Borsboom, D., Perugini, M., Mõttus, R., Waldorp, L. J., & Cramer, A. O. J. (2015). State of the aRt personality research: A tutorial on network analysis of personality data in R. *Journal of Research in Personality, 54*, 13–29.

Epskamp, S., Cramer, A., Waldorp, L., Schmittmann, V. D., & Borsboom, D. (2012). qgraph: Network visualizations of relationships in psychometric data. *Journal of Statistical Software, 48* (4), 1–18. http://www.jstatsoft.org/v48/i04/

Fayyad, U., Piatetsky-Shapiro, G., & Smyth, P. (1996). From data mining to knowledge discovery: An overview. In U. Fayyad, G. Piatetsky-Shapiro, P. Smyth, & R. Uthurusamy (Eds.), *Advantages in knowledge discovery and data mining* (pp. 1–36). MIT Press.

Hair, J. F., Black, W. C., Babin, B. J., & Anderson, R. E. (2014). *Multivariate data analysis* (new int. ed.). Pearson Education.

Hoppock, R. (1935). *Job satisfaction*. Harper.

Latham, G. P., & Budworth, M. H. (2007). The study of work motivation in the 20th century. In L. L. Koppes (Ed.), *Historical perspectives in industrial and organizational psychology* (pp. 353–381). Lawrence Erlbaum.

Locke, E. A. (1970). Job satisfaction and job performance: A theoretical analysis. *Organizational Behavior and Human Performance, 5*(5), 484–500. https://doi.org/10.1016/0030-5073(70)90036-x

Revelle, W. (2018). *Procedures for psychological, psychometric, and personality research*. https://cran.r-project.org/web/packages/psych/psych.pdf

Salkind, N. J. (2010). *Encyclopedia of research design*. SAGE. https://doi.org/10.4135/9781412961288

Scott, J. (2000). *Social network analysis: A handbook* (2nd ed.). SAGE.

Spearman, C. (1904). The proof and measurement of association between two things. *American Journal of Psychology, 15*, 72–101.

Spector, P. E. (1997). *Job satisfaction: Application, assessment, causes and consequences*. SAGE.

Stevens, S. S. (1946). On the theory of scales of measurement. *Science. New Series., 103*(2684), 677–680.

Tang, S., Siu, O., & Cheung, F. (2014). A study of work–family enrichment among Chinese employees: The mediating role between work support and job satisfaction. *Applied Psychology, 63*(1), 130–115.

The Hammond Times. (1957, November 10). Work with new electronic 'Brains' opens field for army math experts p. 65. Retrieved September 20, 2021, from https://www.newspapers.com/clip/50687334/the-times/

Wei, T., Simko, V., Levy, M., Xie, Y., Jin, Y., & Zemla, J. (2017). *corrplot: Visualization of a correlation matrix*. https://cran.r-project.org/web/packages/corrplot/corrplot.pdf

Weiss, H. M., & Merlo, K. L. (2015). Job satisfaction. *International Encyclopedia of the Social & Behavioral Sciences*, 833–838. https://doi.org/10.1016/b978-0-08-097086-8.22029-1

Chapter 3
Fuzzy Cognitive Maps

Fuzzy Cognitive Maps (FCMs) have gained popularity within the scientific community due to their capabilities in modeling and decision making for complex problems. In order to examine and estimate the FCM inference, we develop and the '*fcm*' open-source package R package. Specifically, the *fcm.infer* function provides six inference rules and four threshold functions. This open-source 'fcm' package is available in CRAN and provides the opportunity for everyone to run different scenarios in their weighted matrices. Dedicated examples and visualizations are provided to show the proposed open-source FCM package for business modelling and decision making.

3.1 Introduction

As data modeling is referred to the process of creating a model structure from a dataset that can be useful for recognizing patterns, making decisions and/or predictions and control systems (Brunton & Beyeler, 2019). In other words, data modeling is a visual representation of a problem which often is conceived as a graph (Chapter 4 describes analytically the meaning of data modeling). After model construction, the simulation process is applied in which properties concerning the behavior of the real system or part of the system can be inferred. Specifically, simulation is a technique to evaluate the performance of the system under different circumstances, scenarios or cases (Maria, 1997). In other words, the computer model is a snapshot of a system; whereas simulation consists of consecutive snapshots of the system (that constitute an animated video) describing the behavior of the system after some time/steps (Craiger et al., 1996). In many situations, the model output is compared with the system or experts' output under known input states (Maria, 1997); while, in other cases, simulation involves a systematic study of the behavior of a model (Craiger et al., 1996). Generally, simulation models could be deterministic (input and output are fixed values) or stochastic (at least one of the input or

Z. Dikopoulou, *Modeling and Simulating Complex Business Perceptions*, Fuzzy Management Methods, https://doi.org/10.1007/978-3-030-81496-0_3

output variables is probabilistic); static (time is irrelevant) or dynamic (determining time-varying connections between variables). Most frequently, simulation models are stochastic and dynamic (Maria, 1997).

One of the well-known methods of modeling and simulating of a system based on expert knowledge is the Fuzzy Cognitive Map (FCM). An FCM is a graphical representation of a directed weighted map consisted of concepts and weighted edges. Each concept represents a certain characteristic of the system which is connected to one or more concepts using directed weighted edges, expressing the causal-effect relationships among concepts. This method was primarily proposed by Kosko (1986) for understanding, modeling and simulating systems with numerous interconnections between important components. FCMs adapt the desirable properties of fuzzy logic and neural networks like the ability to represent the structured knowledge of the system and the computation of the inference using a numeric matrix operation instead of straightforward IF-THEN rules (Kosko, 1992).

In recent years, FCMs have been applied in many scientific areas such as social and political sciences, ecological and environmental management, engineering, information technology, robotics, expert systems, medicine, education, neuroscience and so on (Kiani et al., 2019; Amirkhani et al., 2017; Papageorgiou & Salmeron, 2013; Papageorgiou & Kontogianni, 2012). In business modeling, the FCMs have been used in a limited number of applications where the structures of the FCMs were mainly designed from expert knowledge and the literature. At first, Craiger et al. (1996) applied the FCM methodology to model a job-satisfaction problem of 22 variables reflecting personal and organizational aspects. Particularly, in that research, the FCM method and the Structural Equation Method (SEM) were compared. The results have shown that FCM was depending on a small-group of expert knowledge to model a problem which the weights could be changed easily. These changes could be observed to identify the behavior of the system. On the other side, SEM required sufficient data to model a system since small datasets were led to non-convergence of solutions and the inability to estimate parameters. Larger FCM of 165 concepts correlated to business and Information Technology (IT) domain and 210 interconnections were used to simulate the process of Strategic Planning of Information Systems (SISP; Kardaras & Karakostas, 1999). The concepts of the model were derived from the literature and experts' knowledge; thus, the fuzzy weights of the causal relationships were indicated by planners. Different scenarios were developed to access alternative ways of applying IT to improve organizational performance. This procedure allowed planners to simulate and evaluate the impact on the business objectives.

At the beginning of the twenty-first century, the construct and the operation of a reasonable FCM model as a supporting tool to perform a Performance-Driven Change (PDC) activity was presented (Xirogiannis & Glykas, 2004). The structure of the proposed model for Business Progress Reengineering (BPR) was dynamic and hierarchical, including four layers. Every layer included one or more subgraphs of FCMs. The output of every sub-FCM was used to activate another sub-FCM. Moreover, additional inputs were included such as: the business models and strategic BPR plans, including the concepts, the relationship and the structure of the model.

During the process, a re-designer was modifying the causal fuzzy weights among 15 performance concepts. The outputs of the proposed method were crisp values, which compared with experts' reasoning to evaluate the accuracy of the model. Three years after, the authors (Xirogiannis & Glykas, 2007) built a model as a support tool to e-business maturity following the methodology of a previous model for BPR (Xirogiannis & Glykas, 2004). The simulation process of the complex strategic network was applied to quantify the impact in the e-business efficiency. Afterwards, experts were involved to provide the initial values for the concepts and the linguistic values for the causal interconnections.

In 2008, the evaluation of customer satisfaction was conducted in the banking industry in Iran (Nasserzadeh et al., 2008). In total, 31 variables were collected from the literature in which ten experts were determined the sign of each connection (positive, negative or zero) among the concepts. The consensus of these relations was conducted with the Delphi method. Afterwards, 62 experts were used to assign the strength of the causal relationships providing the linguistic weights. As it mentioned in the previous paragraph, Craiger et al. (1996) compared the FCM with the SEM; nevertheless, Lee and Ahn (2009) attempted to incorporate SEM (for modeling) and FCM (for simulation) in a business problem. They collected data from 120 interviewers (employees, managers and executives) to model 19 variables concerning the security and the integrity of business to consumer e-commerce web-based systems (ECWS). First, SEM was applied to identify eight latent variables (the most relevant factors of control design), the strengths of the connections between the latent variables and the directions of the weighted edges. Then, FCM simulation was used to run 'what-if' scenarios to predict strategic performance.

During this decade, Huang et al. (2010) used FCM to model and predict the operation dynamics for third Party Logistics providers (3PLs). In particular, 11 concepts derived from the literature and interviews with experts. Furthermore, human reasoning determined the causal weights among concepts. Then, in scenario analysis proved the notable impact of the significant variables (according to network analysis) to the entire system. In total, they inferred that FCM was a useful tool for quick and comprehensive explorations in which without FCM simulation, the procedure would last days of observation. Recently, a hybrid expert system combining FCM and Analytic Hierarchy Process (AHP) was proposed to describe graphically and monitor Enterprise Content Management (ECM) performance (López & Ishizaka, 2018). Experts in the business domain were used to define the most relevant organizational and technical concepts and their interconnections. Simulation routines were applied to predict the influence of changes in control concepts on system performance. Authors suggested that the proposed ECM model could assist effectively the decisions of managers. Lastly, Karras and Papademetriou (2018) reviewed suitable analytical management computational techniques such as Markov Chains, Game Theory and FCM that were combined in the Business Process Modeling (BPM) framework based on the market competition analysis. Specifically, seven brands of smartphones were used to evaluate the customers' satisfaction which the knowledge of engineers and experts was involved to build the FCM.

Since this dissertation employs various R packages, we develop the '*fcm*' R package as a useful methodological tool for researchers and members of the FCM community to make decisions and policies on their case studies (Dikopoulou & Papageorgiou, 2017). The package is available from the Comprehensive R Archive Network (CRAN) at http://CRAN.r-project.org/. Its main advantages lie in the fact that R is a free software and it provides online help. The fundamental 'fcm' package provides to users a selection of six different inference rules (the main three are: Kosko, modified-Kosko, rescale and the clamping versions which are: clamped-Kosko, clamped-modified-Kosko, clamped-rescale) and four threshold functions (bivalent, trivalent, sigmoid and hyperbolic tangent) to obtain the FCM inference (Kosko, 1986, 1992). The clamping types of these rules are determined when the activated concepts (concept's value equals to 1) are activated in all states during the FCM simulation until the output is stabilized. Step-by-step instructions on how to use an 'fcm' package are provided to acquire the FCM inference of the system related to the job-satisfaction problem.

The rest of the chapter is organized as follows: Section 3.2 introduces the FCM methodology, including the origins of FCM and the simulation process. Next, a step-by-step introduction of our developed R package named 'fcm' is presented in Sect. 3.3. Further recommendations for users of the 'fcm' package are described in Sect. 3.4 and finally, Sect. 3.5 outlines the conclusions.

3.2 Fuzzy Cognitive Maps (FCMs)

FCM is classified as a soft computing technique which is trying to mimic humans' reasoning and decision making. Particularly, an FCM is a combination of fuzzy logic and neural networks, which is used to model and simulate complex systems (Kosko, 1992). Graphically, it is an oriented network consisting of nodes and weighted edges that altogether represent the causal knowledge in a model form. In computer science, it is defined as a fuzzy directed graph with feedback loops.

3.2.1 The History of the Fuzzy Cognitive Maps

The origin of FCM is based on graph theory (Biggs et al., 1976). Later, Axelrod (1976) proposed the Cognitive Maps (CMs) to represent graphically comprehending processes in decision making. The CMs are directed graphs or directed unweighted graphs. The construction of a cognitive map is simple. First, the important variables are defined and then; the causal relationships among the observed variables are specified with arrowheads selecting one of the trivalent values $\{-1, 0, 1\}$. After a decade, Kosko (1986) modified the binary cognitive maps into fuzzy cognitive maps applying fuzzy numbers to the edges that appear cause-effect relations among variables. Since it is difficult for people to assign real numbers for the causal connections to express the strengths of the relationships, it is appropriate to use linguistic variables.

Fuzzy set theory is a generalization of classical set theory in mathematics that limits the set of membership to binary values, indicating that it belongs to a group or not (Zadeh, 1965). In contrast, fuzzy set theory is characterized by grades of membership of the interval −1 to 1, $\mu : U \rightarrow [-1, 1]$. A fuzzy set $\widetilde{A}$ is defined by a membership function within a certain interval, $\mu_{\widetilde{A}}(x) : X \rightarrow [0, 1]$. If $\mu_{\widetilde{A}}(x)$ equals with zero, the element x does not belong to the fuzzy set $\widetilde{A}$. If $\mu_{\widetilde{A}}(x)$ equals with 1, the element x belongs completely to the fuzzy set $\widetilde{A}$. If $\mu_{\widetilde{A}}(x)$ has a value in the range $[0, 1]$, the element x belongs partially to the fuzzy set $\widetilde{A}$ (Zadeh, 1965; Dikopoulou et al., 2017). Therefore, an expert can describe the cause-effect relationships using linguistic variables; for example, the 11 linguistic variables could be determined as: *T(influence} = {negatively very strong, negatively strong, negatively medium, negatively weak, negatively very weak, zero, positively very weak, positively weak, positively medium, positively strong, positively very strong}*. Every linguistic variable corresponds to a membership function in the range [−1, 1], $\mu = \{\mu_{nvs}, \mu_{ns}, \mu_{nm}, \mu_{nw}, \mu_{nvw}, \mu_{z}, \mu_{pvw}, \mu_{pw}, \mu_{pm}, \mu_{ps}, \mu_{pvs}\}$. Lastly, the Centre of Gravity (CoG) defuzzification method (Zadeh, 1986) is used to transform the linguistic variables into crisp numerical values.

FCM, as every graph-based model, consists of two basic components, the nodes and the edges. A set of nodes reflect attributes, characteristics, variables, factors of the system which are known as concepts; while, edges define the causal relationships among two nodes. Each concept is indicated by C_i, $i = 1, 2, \ldots, p$; where n denotes the total number of variables. Furthermore, every concept is carrying a value revealing the degree of activation of the concept in the system at a particular time. These concepts are connected with weighted arcs signifying the cause-effect relationships. Each weight w_{ij} indicates the strength of the association between C_i and C_j taking value in the range −1 to 1. Specifically, w_{ij} measures how much C_i affects C_j ($C_i \rightarrow C_j$). The weights of the map are gathered in the weighted asymmetric adjacency matrix W. There are three possible types of causal relationships among concepts:

- $w_{ij} > 0$ indicates positive causality between concepts C_i and C_j. That is, an increase (decrease) on the value of C_i leads to an increase (decrease) on the value of C_j.
- $w_{ij} < 0$ indicates negative causality between concepts C_i and C_j. That is, an increase (decrease) on the value of C_i leads to a decrease (increase) on the value of C_j.
- $w_{ij} = 0$ indicates no causal relationship between C_i and C_j.

In Fig. 3.1, the procedure of constructing an FCM is presented in four steps. First, the concepts are determined; in this example, there are four concepts, $C_1 - C_4$. Then, experts indicating the causal-effect relations among concepts and next, for every edge, the linguistic variables are indicated. Finally, the linguistic variables are transformed into crisp numerical values.

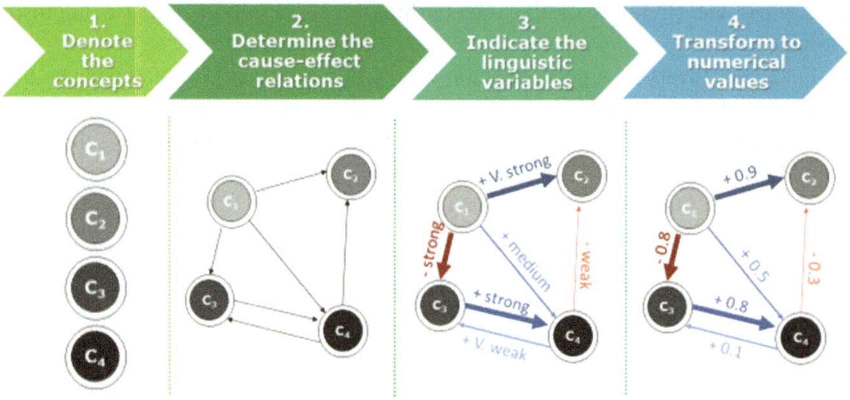

Fig. 3.1 The construction of an FCM in four steps. This FCM includes four concepts and six causal relationships. Blue and red edges denote positive and negative causal relationships, respectively; while, the saturated and wider edges indicate stronger relationships

3.2.2 FCM Inference and Simulation

Kosko first computed the outcome of an FCM known as "FCM inference" applying various policy cases using a computational method adapting from neural network theory (Kosko, 1986). Furthermore, the FCM approach handles humans' perception illustrating how a particular system works. After assigning the initial values for the concepts, they are inserted in the initial numerical $1 \times p$ vector V. The initial vector state at time-step 0 ($\kappa = 0$) of p concepts is denoted as: $V^0 = \left(V_1^0, V_2^0, \ldots, V_p^0\right)$. Each element of the vector represents a value of the concept which it is induced by activating concepts of interest (turning concepts on or off), $V_i^\kappa \in [0, 1]$. Additionally, the Iterative Vector Matrix (IVM) is used to maintain the effects of the concepts at every κ time step.

The initial vector interacts with the connection weighted matrix W. Specifically, W is a squared matrix $p \times p$ which contains all causal w_{ij} weights. The FCM inference rule (5.1) proposed by Kosko (1986) is determined by multiplying the weight adjacency matrix W with the initial state of the vector V (the neural network computational method). This procedure is repeated updating the concepts' values until converges in a steady-state or in other words, it reaches the equilibrium point. Generally, having assigned values V_i^0 and the causal weights w_{ij}, the concepts' values are updating using one of the following calculation rules, (3.1–3.3):

Kosko:

$$\mathrm{V}_\mathrm{i}^{(\kappa+1)} = \mathrm{f}\left(\sum_{\mathrm{j}=1, \mathrm{j}\neq \mathrm{i}}^{\mathrm{n}} \mathrm{w}_{\mathrm{ji}} \times \mathrm{V}_\mathrm{j}^{(\kappa)}\right) \tag{3.1}$$

Modified-kosko:

$$V_i^{(\kappa+1)} = f\left(V_i^{(\kappa)} + \sum_{j=1, j\neq i}^{n} w_{ji} \times V_j^{(\kappa)}\right) \tag{3.2}$$

Rescale:

$$V_i^{(\kappa+1)} = f\left(\left(2 \times V_i^{(\kappa)} - 1\right) + \sum_{j=1, j\neq i}^{n} w_{ji} \times \left(2 \times V_j^{(\kappa)} - 1\right)\right) \tag{3.3}$$

where $V_i^{(\kappa+1)}$ is the value of a concept C_i at simulation step $\kappa + 1$, $V_j^{(\kappa)}$ is the value of a concept C_j at the simulation step κ, w_{ij} is the weight of the interconnection between concept C_i and concept C_j, κ (Greek letter Kappa) is the interaction index at every simulation step. Lastly, $f(\cdot)$ is the threshold or a transformation function that constrains the results of the initial vector V into the range between $[0, 1]$ or $[-1, 1]$, depending on the threshold function. Typically, there are four most commonly used transformation functions (3.4–3.7): (a) bivalent, (b) trivalent, (c) sigmoid and (d) hyperbolic tangent.

Bivalent:

$$f(x) = \begin{cases} 1, & x > 0 \\ 0, & x \leq 0 \end{cases} \tag{3.4}$$

Trivalent:

$$f(x) = \begin{cases} 1, & x > 0 \\ 0, & x = 0 \\ -1, & x < 0 \end{cases} \tag{3.5}$$

Sigmoid:

$$f(x) = \frac{1}{1 + e^{-\lambda x}} \tag{3.6}$$

Hyperbolic tangent:

$$f(x) = \tanh(\lambda \times x) = \frac{e^{\lambda x} - e^{-\lambda x}}{e^{\lambda x} + e^{-\lambda x}} \tag{3.7}$$

where λ is a real positive number ($\lambda \geq 0$), which determines the steepness of the continuous function f and x is the value $V_i^{(\kappa)}$ on the equilibrium point. The FCM model of the system takes the initial values of concepts and weights based on experts' knowledge and experience for the real system and then it is free to interact.

At each step, the value V_i of a concept is influenced by the values of concepts connected to it and it is updated according to the inference rule. This process continues until i) the system converges in a fixed equilibrium point in which the difference between two subsequent values of the outputs must be equal, $V_i^{\kappa} = V_i^{(\kappa+1)}$ or lower to the residual ε (epsilon; in most of the cases is equal to 0.001), $V_i^{\kappa} - V_i^{(\kappa+1)} \leq \varepsilon$; ii) a limited cycle is reached or iii) a chaotic behaviour is revealed. In chaotic behaviour, small changes in the initial stages of the model can lead to significant differences in the following stages (Robertson & Combs, 1995). Below, the simulation of FCM is presented in seven steps:

Algorithm 3.1 The *simulation of FCM*.

1. Indicate the number of steps κ.
2. Initialize the vector V by activating the concepts as scenario demands.
3. Determine the weighted matrix W.
4. Apply one of the inference rules (3.1–3.3) to update the initial Vector V.
5. Use the threshold function (3.4–3.7) to reduce the unbounded input of V to a strict range that threshold controls.
6. Save the updated $V^{(\kappa+1)}$ and consider that it is the initial vector V^{κ} for the next iteration.
7. Repeat steps 4 to 6 until $V_i^{\kappa} = V_i^{(\kappa+1)}$ or $V_i^{\kappa} - V_i^{(\kappa+1)} \leq \varepsilon$ OR $\boldsymbol{\kappa} \leq m$

a. If $\kappa = max$ AND ($V_i^{\kappa} \neq V_i^{(\kappa+1)}$ or $V_i^{\kappa} - V_i^{(\kappa+1)} > \varepsilon$) THEN increase the steps κ in step 1.

In most studies, the sigmoid function is widely applied to obtain the inference of the system (Papageorgiou, 2014; Papageorgiou & Salmeron, 2013). Moreover, in accordance with the interests of the decision-maker, scenario analysis can be accomplished with either all concepts or a subset of concepts being activated depending on the problem. Therefore, different policy scenarios are determined to estimate the inference of the system (Kosko, 1986) answering the "*what—if*" conditions.

3.3 The 'fcm' R Package for FCM Inference Process

Below, stepwise instructions on how to use the **'fcm'** R package are provided (Dikopoulou & Papageorgiou, 2017; Dikopoulou et al., 2018). The function **fcm.infer()** of the 'fcm' R package is applied to estimate the inference of the FCM (Dikopoulou & Papageorgiou, 2017). The function's inputs are two, the initial vector (or else the policy scenario) and the weighted matrix. Additionally, five arguments must be set such as: the number of iterations, the inference rule, the threshold function, the lambda and the epsilon parameter (Fig. 3.2). Thus, the output in the R environment is depicted in Fig. 3.4. The fcm.infer function returns three objects: (a) a data frame which contains the concepts' values of each iteration, (b) an FCM map and (c) the convergence plot of κ steps. In total, six steps are necessary to accomplish the FCM simulations in the R environment which are the following:

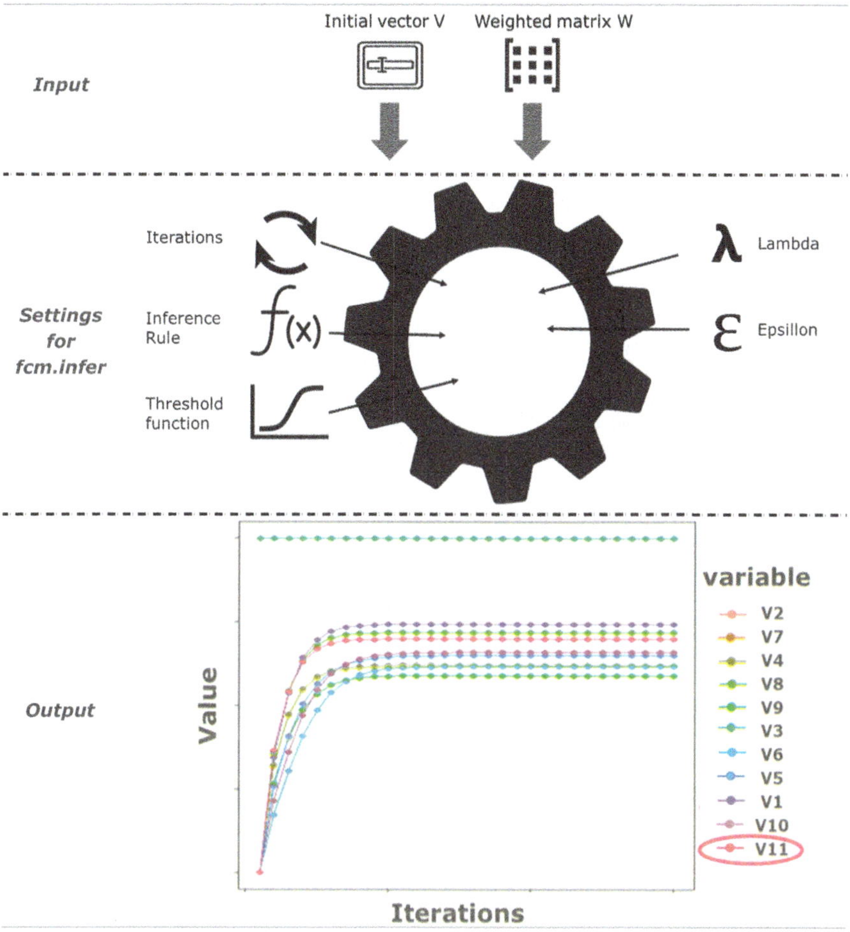

Fig. 3.2 The inputs, the settings and the output of the fcm.infer

Step 1 *Install and load the "fcm" package*: The fcm package can be downloaded and installed only once from the Comprehensive R Archive Network (CRAN; http://CRAN.r-project.org/). Afterwards, the package can be loaded.

```
install.packages("fcm")   # Install the fcm package
library(fcm)         # Load the column names
```

Step 2 *Create the initial activation vector*: In R, a single-dimensional vector is created using the `data.frame` function. From the table in Figure 6.3g, let's assume that five scenarios of the Business sector are selected to be demonstrated. Specifically, the 2nd, the 5th, the 6th, the 12th and the 14th scenario. As it is observed, the 2nd scenario activates only the concept *interesting jobs* ($C_7 = 1$);

the 5th scenario activates the *strong management* ($C_9 = 1$), the 6th scenario triggers two concept ($C_2 = C_7 = 1$), the 12th scenario initiates three concepts ($C_2 = C_7 = C_8 = 1$), and the 14th scenario activates three concepts ($C_2 = C_3 = C_7 = 1$). In order to encode each initial vector as single-dimensional vector in R programming language, we will use the data.frame function. Consequently, the sixth scenario of the table in Figure 8.5g is encoded as V_6 <– data.frame(0, 1, 0, 0, 0, 0, 1, 0, 0, 0, 0). It is important to report that in this example, V_6 indicates the name of the initial vector while V_6 in entire document signifies the name of the variable. The names of the columns of the initial vectors are set with the `colnames()` function.

```
V_2 <- data.frame(0, 1, 0, 0, 0, 0, 0, 0, 0, 0, 0)
V_5 <- data.frame(0, 0, 0, 0, 1, 0, 0, 0, 0, 0, 0)
V_6 <- data.frame(1, 1, 0, 0, 0, 0, 0, 0, 0, 0, 0)
V_12 <- data.frame(1, 1, 0, 1, 0, 0, 0, 0, 0, 0, 0)
V_14 <- data.frame(1, 1, 0, 0, 0, 1, 0, 0, 0, 0, 0)

# Insert the column names of the initial vector V_2
colnames(V_2) <- c("V2", "V7", "V4", "V8", "V9", "V3", "V6",
                   "V5", "V1", "V10", "V11")
# Insert the names of the column V_2 to V_14, V_12, V_6 and
V_5
colnames(V_5) <- colnames(V_6)<- colnames(V_12) <-
                colnames(V_14) <- colnames(V_2)
```

Step 3 *Create the weight matrix*: The weight matrix of Figure 6.3f is created using the `matrix()` function. This matrix (Table 3.1) is the upper triangular part from the output of glasso; while, the weights in the lower triangular matrix are assigned equal to zero. However, the matrix can be asymmetric depending on other studies. To examine how the matrix (or any other object such as data.frame) is looking, `View()` function is applied.

```
W <- matrix(c(0, 0.1207, 0.2499, 0.1617, 0.1183, 0.0514, 0,
 0, 0.3893, 0, 0.0818, 0, 0, 0, 0.3243, 0, 0.1857,
 0.0423, 0.0541, 0.1293, 0, 0.2452, 0, 0, 0, 0.0699,
 -0.0761, 0.0363, 0.2848, 0.2741, 0, 0.0965, 0, 0, 0,
 0, 0, 0.1986, 0.0653, 0, 0, 0.1012, 0.1655, 0, 0, 0,
 0, 0, 0, 0.0836, 0.3772, -0.0647, 0.0636, 0.0993,
 0, 0, 0, 0, 0, 0, 0, 0.0661, 0.1899, 0, 0.1895, 0.117,
 0, 0, 0, 0, 0, 0, 0, 0.0911, 0, 0, 0, 0, 0, 0, 0, 0,
 0, 0, 0, 0, 0.1252, 0, 0, 0, 0, 0, 0, 0, 0, 0, 0, 0,
 0, 0, 0, 0, 0, 0, 0, 0, 0, 0, 0, 0, 0, 0, 0, 0,
 0, 0, 0, 0, 0, 0, 0),nrow=11,ncol=11,byrow=TRUE)
# Add column names in W
colnames(W) <- c("V2", "V7", "V4", "V8", "V9", "V3", "V6",
                 "V5", "V1", "V10", "V11")
View(W)    # View the matrix W
```

Table 3.1 The result of the `View()` function. The weights of the lower triangular matrix are set equal to zero

V2	V7	V4	V8	V9	V3	V6	V5	V1	V10	V11
0	0.121	0.250	0.162	0.118	0.051	0	0	0.389	0	0.082
0	**0**	0	0.324	0	0.186	0.042	0.054	0.129	0	0.245
0	0	**0**	0.070	-0.076	0.036	0.285	0.274	0	0.097	0
0	0	0	**0**	0.199	0.065	0	0	0.101	0.166	0
0	0	0	0	**0**	0.084	0.377	-0.065	0.064	0.099	0
0	0	0	0	0	**0**	0.066	0.190	0	0.190	0.117
0	0	0	0	0	0	**0**	0.091	0	0	0
0	0	0	0	0	0	0	**0**	0	0	0
0	0	0	0	0	0	0	0	**0**	0	0
0	0	0	0	0	0	0	0	0	**0**	0
0	0	0	0	0	0	0	0	0	0	**0**

Alternatively, we can import data (like the weight matrix) into R from different formats (`.csv, .sav, .xlsx, .dta, .systat`) using the `foreign` or the `Hmisc package`.

```
# Import a file
Weight_matrix <- read.table("c:/mydata.csv", header=TRUE)
```

Step 4 Apply the fcm.infer function:

The `fcm.infer()` function is applied to estimate the FCM inference. There are seven arguments (Fig. 3.3) that `fcm.infer()` function needs to initiate for the simulation process.

The initial activation vector `V` is determined as the first argument according to the scenario I to activate the initial state of the concepts,

i) The weighted matrix `W`
ii) The number of iterations.
iii) The selection of inference rule. The third argument must be placed by one of the provided inference rules (Kosko: '*k*', modified Kosko: '*mk*', Rescale: '*r*', Kosko-clamped: '*kc*', modified Kosko-clamped: '*mkc*' or Rescale-clamped: '*rc*').
iv) The transformation function (transform) is specified. Four transformation functions are available (Bivalent: '*b*', Trivalent: '*tr*', Sigmoid: '*s*' or Hyperbolic tangent: '*t*').
v) The parameter lambda (λ) is determined as a positive number, higher than 0 and lower-equal to 10, $0 < \lambda \leq 10$ and
vi) The possible residual value ε (epsilon) is defined between 0.01 and 0.000001. In case that values of λ and ε are not be defined in the fcm.infer() function, the default values for lambda (λ) is 1 and for *e* (ε, epsilon) is 0.001.

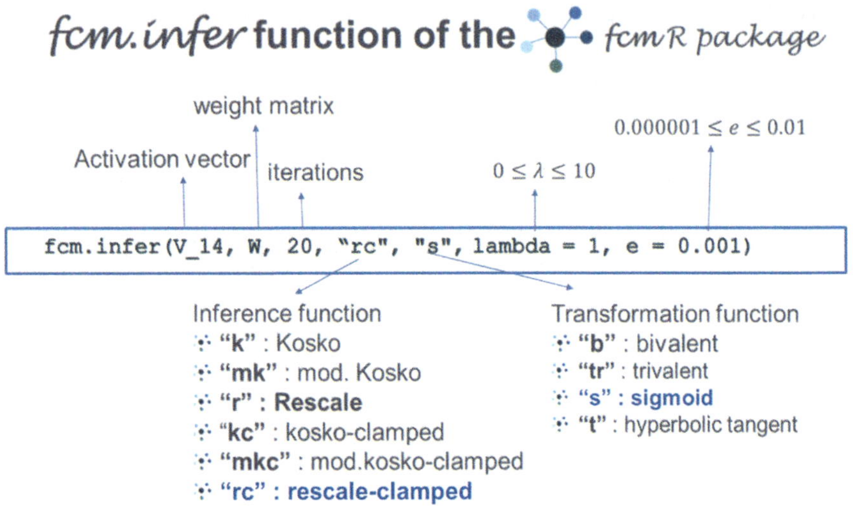

Fig. 3.3 The fcm.infer function to simulate the observed job-satisfaction network. The `fcm.infer()` function requires seven arguments of the: activation vector, weight matrix, iteration, inference function, transformation (threshold) function, lambda and residual ε

Moreover, the '`fcm`' package imports three additional packages: qgraph (Epskamp et al., 2012), reshape (Wickham, 2007), ggplot2 (Wickham, 2016) which are useful for creating elegant graphs and plots. This indicates that after the inference estimation, the `fcm.infer` function returns not only a data frame which contains the values of the concepts of each iteration but also graphical objects, like the FCM map and the convergence plot of κ steps.

For the estimation of the FCM inference regarding the five scenarios, they require five lines of coding as the demonstrated scenarios. The inference rule is modified into '*rc*' for rescale clumped rule, Eq. (3.3). The weight matrix (W), the number of iterations (*iter* = 20), the transformation function (Sigmoid '*s*'), *lambda* = 1 and *e* = 0.001 remain the same:

```
# Rescale clamped inference rule
fcm.infer(V_2, W, 20, 'rc', 's', lambda = 1, e = 0.001)
fcm.infer(V_5, W, 20, 'rc', 's', lambda = 1, e = 0.001)
fcm.infer(V_6, W, 20, 'rc', 's', lambda = 1, e = 0.001)
fcm.infer(V_12, W, 20, 'rc', 's', lambda = 1, e = 0.001)
fcm.infer(V_14, W, 20, 'rc', 's', lambda = 1, e = 0.001)
```

Step 5 R*ead the results fcm.infer function*:

The results of the 14th scenario (Figure 6.3g) using the fcm.infer function is presented. It is important to mention that the fcm.infer function stops when the equilibrium point is converged. If the system does not reach the equilibrium point, then a message informs the user to increase the number of iterations to reach the

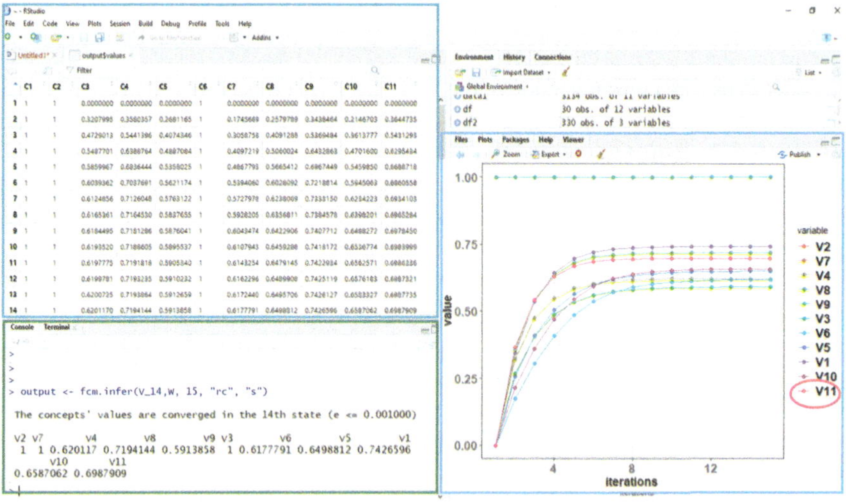

Fig. 3.4 The R environment. The source panel (up left) presents the values of the concepts in every iteration step. The console panel (down left) inform the user in which step the system reaches the equilibrium point and, in the plot/help panel (down right), the plot of the converged values is depicted. The horizontal axis (X-axis) represents the iterations of the FCM simulation; while, the vertical axis (Y-axis) points out the values of the concepts at every simulation step

convergence. Moreover, the converged values of all κ steps appear in the R-console and the convergence plot in the R-plot pane. Specifically, Fig. 3.4 displays the environment of RStudio and the results that the fcm.infer function returns applying the 14th scenario. In the console pane (down-left), the converged values of the vector V_14 are shown. As observed, the iterations stopped at the 14th step and the converged vector (final state) is V^{14}=1, 1, 0.620, 0.719, 0.591, 1, 0.618, 0.650, 0.743, 0.659, **0.699**]. Furthermore, the plot pane (down-right) illustrates graphically the convergent values of the console pane.

The explanation of the simulation is simple. The 14th scenario examines the impact of "Prospects/career opportunities (C_2)", "Pleasant working atmosphere (C_3)" and "Interesting jobs (C_7)" to the output concept of the system which is "Willingness to work (C_{11})". In other words, the policy scenario answers the question "If the Career opportunities will increase and at the same time, Pleasant working atmosphere will be improved and the job will be more interesting ($C_2 = C_3 = C_7 = 1$) what will be the influence on the willingness of work in the FCM?". The activation levels at the final vector and changes/trends in the activation levels throughout the simulation in Fig. 3.4 indicate that the willingness of the employees to work (C_{11}) will be increased (pink line).

3.4 Recommendation to the 'fcm' Users

In order to use the current version of the 'fcm' package (version 0.1.3), it is necessary to install it as described in Sect. 3.1. Ensure that have a recent version of R (at least version 3.3.2) from *https://cran.r-project.org/*. Moreover, it is important to download the latest version of the package you are interested in because bugs have been solved or some extensions have been added. For reporting bugs in 'fcm' package, please submit them in *https://github.com/LiaDD/Fuzzy-Cognitive-Maps-FCMs*.

Concerning the visualizations, every user can modify the colors of the plot according to the suggestions of ggplot2 package which is not the subject of this paperwork. The 'fcm' package includes the standard palette of colors of the ggplot2 as depicted in Fig. 3.2. Furthermore, if the number of concepts is increasing, the execution time of the algorithm is increasing likewise. In other words, the R code will run slower compared to other programming languages (such as Python, C++, etc.) but it will deliver the appropriate results in a more user-friendly environment. According to Wickham (2014), a Chief Scientist at RStudio "*R is not a fast language. This is not an accident. R was purposely designed to make data analysis and statistics easier for you to do. It was not designed to make life easier for your computer. While R is slow compared to other programming languages, for most purposes, it's fast enough*".

3.5 Conclusions

This chapter has presented a well-known method, the Fuzzy Cognitive Map (FCM) which was used in various scientific fields to model and simulate complex systems. Generally, the FCM methodology combines the desirable properties of fuzzy logic and neural networks; subsequently, it is classified as soft a computing technique (Kosko, 1986, 1992; Papageorgiou & Salmeron, 2013). According to the literature, most of the studies were used experts' knowledge or historical data to model the observed problem and then, simulate the model regarding the scenario policy (Dikopoulou et al., 2020a, b).

Since, in this research, we use exclusively the R programming language, we observed that there was not any developed R package corresponding to the FCM methodology. For this reason, we developed the 'fcm' R package, which is available in CRAN (https://cran.r-project.org/web/packages/fcm/) with a variety of adjustments. Particularly, six inference rules, six threshold functions and a range of values for λ and ε parameters are supported. Additionally, we provided helpful information about how to use the 'fcm' package to infer the decision policies on the FCMs. The user first specifies the activation vector and the weighted matrix. Therefore, the fcm.infer() function is used to accomplish the FCM simulation which includes seven arguments as input to the function. Normally, the algorithm of the fcm.infer function stops when the equilibrium point for each concept is reached. Specifically, the values of each concept of each iteration are presented as a graphical plot, writing only a single line of code! In future, we hope more researchers will use this open-source '*fcm*' R package to simulate their decision policies on their fuzzy cognitive maps.

References

Amirkhani, A., Papageorgiou, E. I., Mohseni, A., & Mosavi, M. R. (2017). A review of fuzzy cognitive maps in medicine: Taxonomy, methods, and applications. *Computer Methods and Programs in Biomedicine, 142*, 129–145.

Axelrod, R. (1976). *Structure of decision: The cognitive maps of political elites*. Princeton University Press.

Biggs, N. L., Llooyd, E. K., & Wilson, R. J. (1976). *Graph theory* (pp. 1736–1936). Clarendon Press.

Brunton, B. W., & Beyeler, M. (2019). Data-driven models in human neuroscience and neuroengineering. *Current Opinion in Neurobiology, 58*, 21–29. https://doi.org/10.1016/j.conb.2019.06.008

Craiger, J. P., Goodman, D. F., Weiss, R. J., & Butler, A. (1996). Modeling organizational behavior with Fuzzy Cognitive Maps. *International Journal of Computational Intelligence and Organisation, 1*, 120–123.

Dikopoulou, Z., Papageorgiou, E. (2017). *Inference of Fuzzy Cognitive Maps (FCMs)*. The Comprehensive R Archive Network (CRAN), R package Version 0.1.3. https://cran.r-project.org/web/packages/fcm/vignettes/vignettes.html

Dikopoulou, Z., Papageorgiou, E., Jetter, A., & Bochtis, D. (2018). *Open-source tool in R language to estimate the inference of the Fuzzy Cognitive Map in environmental decision making*. 9th International Congress on Environmental Modeling and Software, Fort Collins, USA, June 24–28, 2018.

Dikopoulou, Z., Papageorgiou, E., & Vanhoof, K. (2020a). Retrieving sparser fuzzy cognitive maps directly from categorical ordinal dataset using graphical lasso models and the max-threshold algorithm. IEEE International Conference on Fuzzy Systems (FUZZ-IEEE).

Dikopoulou, Z., Papageorgiou, E., & Vanhoof, K. (2020b). From undirected structures to directed graphical lasso fuzzy cognitive maps using ranking-based approaches. IEEE International Conference on Fuzzy Systems (FUZZ-IEEE).

Dikopoulou, Z., Papageorgiou, E., Nápoles, G., & Vanhoof, K. (2017). A modified fuzzy TOPSIS method aggregating 8.921 partial rankings for companies' attractiveness. *Fuzzy Management Methods*, 59–71. https://doi.org/10.1007/978-3-319-54048-1_6

Epskamp, S., Cramer, A., Waldorp, L., Schmittmann, V. D., & Borsboom, D. (2012). qgraph: Network visualizations of relationships in psychometric data. *Journal of Statistical Software, 48* (4), 1–18. http://www.jstatsoft.org/v48/i04/

Huang, Y. K., Feng, C. M., Yeh, W. C., & Lin, L. Y. (2010). *A fuzzy cognitive map modelling to explore the operation dynamics of third-party logistics providers*. 2010 International Conference on Logistics Systems and Intelligent Management (ICLSIM). https://doi.org/10.1109/iclsim.2010.5461166

Kardaras, D., & Karakostas, B. (1999). The use of fuzzy cognitive maps to simulate the information systems strategic planning process. *Information and Software Technology, 41*(4), 197–210. https://doi.org/10.1016/s0950-5849(98)00125-6

Karras, D. A., & Papademetriou, R. (2018). A systematic review of analytical management techniques applied to competition analysis modelling towards a framework for integrating them with BPM. In B. Shishkov (Ed.), Business Modeling and Software Design – 7th International Symposium, BMSD 2017, Revised Selected Papers: 7th International Symposium, BMSD 2017, Barcelona, Spain, July 3–5, 2017, Revised Selected Papers (Vol. 309, pp.

166–185). [9] (Lecture Notes on Business Information Processing; Vol. 309). Springer. https://doi.org/10.1007/978-3-319-78428-1_9

Kiani, M.,Andreu-Perez, J.,Hagras, H., Papageorgiou, E. I., Prasad, M., & Lin, C. (2019). *Effective brain connectivity for fNIRS with fuzzy cognitive maps in neuroergonomics*. IEEE Transactions on Cognitive and Developmental Systems.

Kosko, B. (1986). Fuzzy cognitive maps. *International Journal on Man-Machine Studies, 24*, 65–75.

Kosko, B. (1992). *Neural networks and fuzzy systems: A dynamical systems approach to machine intelligence*. Prentice-Hall.

Lee, S., & Ahn, H. (2009). Fuzzy cognitive map based on structural equation modelling for the design of controls in business-to-consumer e-commerce web-based systems. *Expert Systems with Applications, 36*(7), 10447–10460. https://doi.org/10.1016/j.eswa.2009.01.070

López, C., & Ishizaka, A. (2018). A scenario-based modelling method for controlling ECM performance. *Expert Systems with Applications, 97*, 253–265. https://doi.org/10.1016/j.eswa.2017.12.024

Maria, A. (1997). *Introduction to modelling and simulation*. Proceedings of the 29th Conference on Winter Simulation – WSC '97. https://doi.org/10.1145/268437.268440

Nasserzadeh, S. M. R., Jafarzadeh, M. H., Mansouri, T., & Sohrabi, B. (2008). Customer satisfaction fuzzy cognitive map in banking industry. *Communications of the IBIMA, 2*(21), 151–162.

Papageorgiou, E. I., & Kontogianni, A. (2012). Using fuzzy cognitive mapping in environmental decision making and management: a methodological primer and an application. In S. S. Young & S. E. Silvern (Eds.), *International perspectives on global environmental change* (pp. 427–450). InTech.

Papageorgiou, E. I., & Salmeron, J. L. (2013). A review of fuzzy cognitive maps research during the last decade. *Fuzzy Systems, IEEE Transactions, 21*, 66–79.

Papageorgiou, E. I. (2014). Fuzzy cognitive maps for applied sciences and engineering from fundamentals to extensions and learning algorithms. *Intelligent Systems Reference Library, 54*.

Robertson, R., & Combs, A. (Eds.). (1995). *Chaos theory in psychology and the life sciences*. Lawrence Erlbaum Associates.

Wickham, H. (2014). *Advanced R*. Chapman and Hall/CRC.

Wickham, H. (2007). Reshaping data with the reshape package. *Journal of Statistical Software, 21* (12).

Wickham, H. (2016). *ggplot2: Elegant graphics for data analysis*. Springer.

Xirogiannis, G., & Glykas, M. (2004). Fuzzy cognitive maps in business analysis and performance-driven change. *IEEE Transactions on Engineering Management, 51*(3), 334–351. https://doi.org/10.1109/tem.2004.830861

Xirogiannis, G., & Glykas, M. (2007). Intelligent modelling of e-business maturity. *Expert Systems with Applications, 32*(2), 687–702. https://doi.org/10.1016/j.eswa.2006.01.042

Zadeh, L. A. (1965). Fuzzy sets. *Information and Control, 8*(3), 338–353. https://doi.org/10.1016/s0019-9958(65)90241-x

Zadeh, L. A. (1986). A simple view of the Dempster-Shafer theory of evidence and its implication for the rule of combination. *AI Magazine, 7*(2), 85. https://doi.org/10.1609/aimag.v7i2.542

Chapter 4
Data Modeling

Over recent years, researchers investigate methods to model data-driven problems from cross-sectional data as a causal view of latent factors. Nevertheless, some real complex problems could be efficiently represented as parsimonious and interpretable network structures which the direct interactions among the observed variables are estimated. This dissertation adopts the methodology of regularized partial correlation networks; a well-known method originated from the field of machine learning to model a business data-driven problem. To the best of our knowledge, the proposed methodology is applied for the first time in a business field to estimate network structures. In total, 56,246 Belgian employees working in 15 sectors evaluate ten job-satisfaction variables. Comparisons with the previous known methodology for creating network constructions based on the distance-based algorithm show that the proposed approach is faster (particularly in large datasets). Moreover, it provides more understandable/useful relationships among nodes, through a less complex structure for making decisions. All analyses are performed using the R programming language.

4.1 Introduction

Data modeling is the process by which the data are organized to represent as closely as possible a real-world problem (Vazigiannis, 2003). In the past, exploratory studies such as Factor Analysis (FA) were applied to describe the relationships between the observed variables with a smaller number of unobserved variables, named as "factors" (Thompson, 2007). In other words, Factor Analysis aims to identify any underlying factors that are responsible to explain the structural covariation between a set of independent variables. Moreover, it is characterized as a multivariate procedure which mostly applied in social science (e.g., Mike et al., 2019; Grover et al., 2018; Norris & Lecavalier, 2009). Originally, FA was introduced in psychological theory and was officially termed as "factor analysis" by Thurstone in 1931.

Z. Dikopoulou, *Modeling and Simulating Complex Business Perceptions*, Fuzzy Management Methods, https://doi.org/10.1007/978-3-030-81496-0_4

The FA modeling was utilized on cross-sectional data (Poghosyan et al., 2009; Dahlin et al., 2005; Loh & Venkatraman, 1992). A cross-sectional study is defined as an observational study that can be conducted with different samples, including multiple variables that can be studied at a given point in time (Knowland et al., 2015). However, cross-sectional data may not provide definite information about cause-effect relationships, but they identify associations among the observed variables. The reason is that these data provide a snapshot of a single moment in time and do not have any information about the time before or after the snapshot is taken (Levin, 2006).

Nevertheless, not all complex problems can be explained as a causal view of latent factors (Costantini et al., 2015; Cramer et al., 2012). In many real problems, a common cause can acquire with difficulty independent associations with its symptoms that explain their emergence and covariance (McNally, 2016). Some problems are more suited to be represented as a network which models the most relevant and direct relations among the observed variables.

A network is an abstract model which represents complex phenomena such as technological and biological structures, information networks, social relations (Newman, 2010). Particularly, networks consist of two components, nodes and edges. Nodes or vertices (V) represent entities (variables, attributes, objects, people, etc.); while, edges or links (E) connect the nodes between them representing their connections, which are generally observed and known (such as friendships, electricity lines, citations and so on). A set of tuples represents the edges, in which each tuple determines a pair of nodes that are existing on that edge. In mathematical terms, a network is called as a "graph" and it explains a detailed description of network structure). However, in psychological networks (Epskamp et al., 2017a, b; Epskamp & Fried, 2017), the strength of connection among two nodes in unknown and it is estimated from data. In fact, a psychological network is comprised of the basic components of normal network (nodes and edges), but edges represent statistical relationships between nodes.

In recent years, the most common method to estimate a psychological network is the pairwise Markov Random Fields (MRF; Epskamp et al., 2018; Epskamp & Fried, 2017; Murphy, 2012). When the observed data are continuous and normally distributed, the derived models are classified under the Gaussian Graphical Model (GGM) family (Lauritzen, 1996). In GGM, the estimated edges represent the partial correlation coefficients between two variables after conditioning on all other variables in the dataset. These GGM networks are known as partial correlation networks which usually estimated using regularization techniques originated from the field of machine learning (Murphy, 2012). Prior research generally confirms that regularization performs well in estimating the true network structure (Meinshausen & Bühlmann, 2006; Friedman et al., 2008; Foygel & Drton, 2010; Costantini et al., 2015; Epskamp et al., 2017a, b; Epskamp & Fried, 2017).

Regularization is the most convenient process to avoid overfitting by regularizing or shrinking many estimated coefficients exactly to zero. This signifies that spurious edges are removed from the model, indicating that model interpretability will be increased due to the simplicity of the estimated networks. The regularized networks have been applied widely to psychological data such as clinical and psychology, psychiatry, personality social research and health sciences (e.g., De Ron et al., 2019; Isvoranu et al., 2019; Epskamp et al., 2017a, b; Epskamp & Fried, 2017; Fried et al., 2016; McNally 2016; Costantini et al., 2015; van Borkulo et al., 2015; Kossakowski et al., 2015). To the best of our knowledge, this method has not yet been introduced in business literature. In this study, we model cross-sectional ordinal datasets to investigate the connectivity between the certain variables which are related to the job-satisfaction problem.

The aims of this chapter are: 1) to describe a network and classify the types of networks, 2) to provide an introduction to regularized partial correlation networks, 3) to estimate statistical models on cross-sectional data (the rows of the data are independent) concerning the job-attractiveness problem (business networks). Particularly, we investigate the pairwise connections between 10 job-satisfaction factors of 15 different working sectors. Thus, in this chapter we 4) visualize the parameters of each model as a weighted network, 5) compare regularized networks to distanced-based methods (Schneider et al., 1998) and 6) examine similarities and differences between the regularized networks.

4.2 Classification of Networks

Generally, any graph G consists of a vertex set $V = \{1, 2, \ldots, p\}$ and an edge set $E \subseteq V \times V$. A graph $G = (V, E)$ may be defined as either undirected or directed according to the connectivity of the nodes which characterize the structure of the network. In addition, edges may be either determined as unweighted or weighted. The beneficial representation of a network is determined by $p \times p$ matrix, where p indicated the number of nodes. In other words, nodes are represented by the columns and rows of the matrix. The relation between two nodes i and j is denoted by a matrix element (i, j).

Graphs can not only be classified according to the type of edges, directed or not but also can be categorized if the edges carrying weights or not. Unweighted networks describe only the presence or absence of edges, Table 4.1. Therefore, the corresponding matrix of an undirected network is called an adjacency matrix in which the relationships between nodes are encoded either 1 or 0 (as a binary form). If

Table 4.1 (a) Reports the symmetric adjacency matrix (left) for the undirected network and (b) displays the asymmetric adjacency matrix (right), respectively

Adjacency matrices										
	(a) Undirected network					(b) Directed network				
	A	B	C	D	E	A	B	C	D	E
A	0	1	1	1	0	0	1	1	1	0
B	1	0	1	0	0	0	0	1	0	0
C	1	1	0	1	1	0	0	0	1	0
D	1	0	1	0	0	0	0	0	0	0
E	0	0	1	0	0	0	0	1	0	0

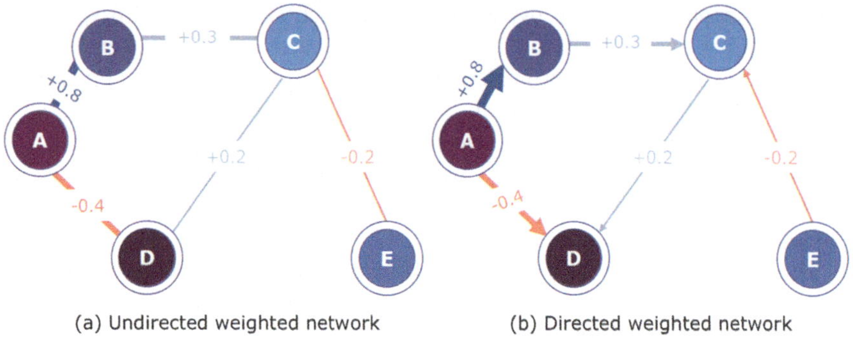

Fig. 4.1 A network of five nodes and six edges. Each edge carries a weight indicating the strength of influence among nodes. Blue edges specify positive relations; while, red edges denote negative relations between nodes. (**a**) illustrates an undirected weighted network (left) and (**b**) illustrates a directed weighted network (right)

Table 4.2 (a) Represents the symmetric weight matrix (left) for the undirected network and (b) displays the asymmetric weight matrix (right) for the directed network of Fig. 4.1a, b, respectively

Weight adjacency matrices										
(a) Undirected weighted network						(b) Directed weighted network				
	A	B	C	D	E	A	B	C	D	E
A	0	0.8	0.1	−0.4	0	0	0.8	0.1	−0.4	0
B	0.8	0	0.3	0	0	0	0	0.3	0	0
C	0.1	0.3	0	0.2	−0.2	0	0	0	0.2	0
D	−0.4	0	0.2	0	0	0	0	0	0	0
E	0	0	−0.2	0	0	0	0	−0.2	0	0

there is a link, then it is defined as 1; otherwise, as 0, indicating the lack of connectivity between two nodes.

On the other hand, weighted networks comprise additional information about the magnitude of the connections. Particularly, a weighted network could be encoded in a weighted matrix which its elements represent the connections' strength among two nodes. As strength is defined as the absolute value of the edge weight; while, the sign, positive or negative, defines the type of the influence. An example of a simple network of five nodes and six edges is depicted in Fig. 4.1. The nodes represent the entities; while the edges represent the relationship between the nodes (Newman, 2010).

For instance, the networks in Fig. 4.1 can be presented as weight matrices. It is noticed that the matrix in Table 4.2a represents an undirected network which is symmetric and the diagonal values are set equal to zero. In addition, the elements (cells) of the weight matrix specifying the relationships among a pair of nodes can be positive and negative. Particularly, there are four positive connections ($A - B$, $A - C$, $C - D$ and $B - C$) and two negative connections ($A - D$ and

$C - E$). It is also observed that there are no interactions among some pairs of nodes, such as node B and E. Accordingly, Table 4.2b shows an asymmetric weight matrix corresponding to the directed network in Fig. 4.1b. Even if the number of positive or negative connections is the same, the interpretation is different. For example, node A can influence positively node B, ($A \rightarrow B$) but node B has no influence to node A.

Moreover, every weight matrix can be encoded (as described above) into an adjacency matrix to justify the presence or the absence of the edge where the weights of the edges are converted equal to value 1 and value 0, otherwise.

4.3 Modelling Networks from Cross-Sectional Data

In recent years, researchers usually do not have any knowledge about the true network structure; consequently, all decision processes and models are estimated from data. This subchapter introduces two methods that manipulate ordinal cross-sectional data, the distanced-based method (Schneider et al., 1998) and Gaussian graphical models (Lauritzen, 1996). Both methods provide indirect symmetric matrices, which are visualized and compared. A visual inspection projects the pairwise interactions between the job-satisfaction factors (variables) to examine not only the plausible interpretation of these relations but also to reveal which variables are most closely related to each other.

4.3.1 *The Distance-Based Method*

In 1998, Schneider et al. introduced the distanced-based method to model a dataset of 18 variables associated with an economic/demographic problem from 106 countries (Schneider et al., 1998). This approach consists of four steps, calculating matrices in each one of the steps, including the Initial Influence Matrix (IIM), the Fuzzified Influence Matrix (FIM), the Strength of Influence Relationships Matrix (SIRM), and the Final Matrix of Influence (FMI). Moreover, all these steps are employed to provide the symmetric weighted matrix of the graph. To conclude the advantage of this method is the simplicity of the model implementation in small datasets. Below, the steps of the distance-based method are described to calculate the structure of the model:

Initial Influence Matrix (IIM): The IIM contains a matrix [p $\times$ m], where p is defined as the number of variables and m the number of interviewees. Each element in this matrix (O_{ij}) represents the importance of the variable i over the individual's j perspective. Hence, a vector V_i denoted as a sequence of data elements O_{i1}, O_{i2},. . ., O_{im} and showed the opinion of an expert $j = 1, \ldots, m$ about different variables i.

Fuzzified Influence Matrix (FIM): Each numerical vector V_i is transformed into a fuzzy set, in the interval [0, 1]. The maximum value in V_i is assigned as: $X_i = 1$, $Max(O_{iq}) \rightarrow X_i(O_{iq}) = 1$. The minimum value in V_i is assigned as $X_i = 0$, $Min(O_{ip}) \rightarrow X_i(O_{ip}) = 0$. The other elements of the vector V_i are normalized in the [0, 1] interval proportionally (Eq. 4.1). The $X_i(O_{ij})$ is expressed as the degree of membership (normalization) of the element O_{ij} to the vector V_i.

$$X_i(O_{ij}) = \frac{O_{ij} - Min(O_{ip})}{Max(O_{iq}) - MinO_{ip}} \tag{4.1}$$

In order to represent the membership grades in a more reasonable way, reflecting the real-world problem, threshold values, upper (a_{up}) and lower threshold (a_{low}) are introduced. Schneider et al. proposed threshold values equal to $a_{up} = 80$ and $a_{low} = 20$, in order to consider a possible deviation of ±20% between the interviewers. In other words, the O_{ij} (in the IIM matrix) were set to a scale from 0 to 100 then values higher than 80 were transformed to 1 and lower to 20 as 0 (Schneider et al., 1998; Rodriguez-Repiso et al., 2007). In this study, the questionnaires were answered in a five-point Likert scale, (1 signifies the lower and 5 the higher preference value). For this reason, following the deviation of ±20%, values equal and higher than 4 (applying the cross-multiplication method $\frac{5*80}{100}$) were transformed to 1 and equal to 1 ($\frac{5*20}{100}$) as 0 in the FMI matrix.

Strength of Influence Relationships Matrix (SIRM): The SIRM is an [p × p] matrix. Each element S_{ij} represents the correlation of factors i and j, in the interval [−1, 1]. To obtain the SIMS matrix, first, the distance d_j and the inverse distance $d_{j\,(inv)}$ (Eq. 4.2) are estimated. Then, the average distances $\overline{D}$ and the average inverse distances $\overline{D}_{inv}$ between all the pairs of variables (Eq. 4.3) are calculated:

$$d_j = |X_1(v_j) - X_2(v_j)| \;\; and \;\; d_{j(\text{inv})} = |X_1(v_j)(1 - X_2(v_j))| \tag{4.2}$$

$$\overline{D} = \frac{\sum_{j=1} \mathrm{d}_j}{m} \text{ and } \overline{D}_{\text{inv}} = \frac{\sum_{j=1} \mathrm{d}_{j(\text{inv})}}{m} \tag{4.3}$$

Next, the similarities between two vectors are calculated and compared.

$$S = 1 - \overline{D} \text{ and } S_{inv} = 1 - \overline{D}_{inv} \tag{4.4}$$

Therefore, there produced three possible relations among two concepts i and j (S_{ij}):

The positive causality: When $S > S_{inv}$, then the value S_{ij} of the SIMS matrix defined as the positive value of S.

The negative causality: When $S < S_{inv}$, then the value S_{ij} defined as the negative value of S_{inv}.

No causality: When $S = S_{inv} = 0$, then the value S_{ij} is equal with zero.

Final Matrix of Influence (FMI): Occasionally, due to meaningless cause and effect correlation between some pairs of concepts, expert's intervention is required to analyze the element S_{ij} of SIRM matrix. According to experts' knowledge, the irrelevant relations could be easily identified and removed (Schneider et al., 1998; Rodriguez-Repiso et al., 2007; Bakhtavar & Shirvand, 2018; Zarrin & Azadeh, 2018). Finally, the FMI matrix determines the relationships among the concepts of the graph according to expert knowledge.

Consequently, this method is categorized as a semi-automated method since limited human intervention is required (Dikopoulou et al., 2017). To conclude, the distance-based method has been applied to model FCMs using small (up to 150 observations) datasets (Ahmadi et al., 2015; Avran et al., 2016; Zarrin & Azadeh, 2018; Bakhtavar & Shirvand, 2018; Stylios et al., 2020). However, this dissertation will create FCM structures from larger (up to 20,300 observations) datasets.

4.3.2 Model Selection Using Graphical Models

Graphical models depended on graph theory and probability theory to create networks that learn complex probabilistic relationships. Hence, the properties of an underlying graph indicate the conditional independence structure between random variables (Alsberg, 2000).

A correlation network is a simple heuristic model to infer a network of p variables. In this case, if two variables X and Y are correlated ($X \sim Y$), then the corresponding nodes are adjacent to each other by an edge. Alternatively, the absence of an edge indicates the absence of a correlation between two nodes and it is denoted with a zero value. However, such pairwise correlation networks tend to characterize as spurious, representing relationships that are not true in reality (Costantini et al., 2015).

A graphical model is applied when the structure of a graph is unknown, eliminating spurious/false relationships and improving the interpretation of the connections between nodes (Hastie et al., 2009). This instigates the usage of Markov networks, widely acknowledged as Gaussian graphical models for Gaussian distributed data (Friedman et al., 2008). The estimated model is based on samples; a problem which is known as a graphical model selection. Other important challenges of graphical models are the estimation of the edge weights from data, which in computer science is called *learning* and the computation of marginal node probabilities from their distributions is known as *inference* (Murphy, 2012).

Before we examine the case of multivariate Gaussians, we briefly introduce the Markov networks for general probability distributions. Then, we demonstrate how Gaussian graphical models are estimated using the inverse covariance matrix.

4.3.2.1 Markov Random Graphs

The main objective of a graphical model is to use the clique structure or the cut-set structure of the underlying graph to constrain the distribution of the random vector $X = (X_1, X_2, \ldots, X_p)$. Any p-dimensional random variable X that has some probability distribution $X \sim \mathbb{P}_X$ can be associated with a vertex set $V = \{1, 2, \ldots, p\}$ of the underlying graph (Hastie et al., 2015). This subchapter describes the basic properties of the indirect graphical models as graphs using joint distributions of a set of random variables, which are often referred to as Markov Random Fields or Markov Networks.

Let us assume that a set of random variables in a graph $\mathcal{G}$ have a joint distribution P. In a Markov graph $\mathcal{G}$, the absence of an edge denotes the conditional independence between random variables given the rest variables of the graph $\mathcal{G}$. The symbol $\perp\!\!\!\perp$ indicates the relation "is conditionally independent of" and symbol $|$ denotes "conditioning (given) by" (Hastie et al., 2015). For example, in Fig. 4.1a, the *pairwise Markov independency* between node D and B of $\mathcal{G}$ is signified by $D \perp\!\!\!\perp B \mid C$; indicating that nodes D and B are conditionally independent given node C.

In graph theory, another significant notation is a subgraph, $U \in V$. It is defined as a subset of vertices including their edges (Hastie et al., 2009). In the graph $\mathcal{G}$ (Fig. 4.1a), if $\mathcal{A} = \{A, B\}$, $\mathcal{B} = \{E\}$ and $\mathcal{S}$={C, D} are subgraphs, then $\mathcal{S}$ is called separator because it separates the graph into two conditionally independent segments $\mathcal{A}$ and $\mathcal{B}$. This relation is determined as $\mathcal{A} \perp \mathcal{B} \mid \mathcal{S}$ and it is known as the *global Markov property* of $\mathcal{G}$. Hence, global Markov property decomposes graphs into smaller and more manageable components driving to plain computation and interpretation. Each of these properties makes an important contribution to our understanding of the equivalence between the *pairwise* and the *global Markov properties* of a graph with positive distributions (Hastie et al., 2009, 2015). Consequently, this is relevant for inferring global independence relations from simple pairwise properties.

The *factorization property* is defined directly in terms of the probability distribution of the random vector X. A fully-connected subset (or complete subgraph) of the vertex set, $C \subseteq VC \subseteq$ V such that $s, t \in E$ for all $s, t \in C$, while $s \neq ts \neq$ t, is called *graph clique.* A clique is considered to be maximal if does not include any other clique. For example, the maximal cliques in the graph of Fig. 4.1a are {A, B},{C, D},{B, C},{A, D},{C, E}. Likewise, the set of cliques (maximal and non-maximal) of the graph $\mathcal{G}$ is denoted as $\mathfrak{C}$C. For each clique C, $C \in \mathfrak{C}$C $\in$ CC $\in$ C, a clique compatibility function $\psi_c(Xc)\psi_C(X_C)$ is a mapping configuration $Xc := \{X_s, s \in V\}$ $X_C := \{X_s, s \in V\}$to $\mathbb{R}^+$ such that ψ_c $\psi_C$only depends on the node X_c corresponding to the clique C. The *Factorization property* is considered by a given set of compatibility functions which the probability distribution $\mathbb{P}$ factorizes over $\mathcal{G}$ if it may be represented as a product of clique functions; where Z is the partition function given by the sum $Z = \sum_{x \in X} \prod_{C \in \mathfrak{C}} \psi_C(X_C)$.

$$\mathbb{P}(X_1, \ldots, X_p) = \frac{1}{Z} \prod_{C \in \mathfrak{C}} \psi_C(X_C) \tag{4.5}$$

The Hammersley–Clifford theorem (Hammersley & Clifford, 1971; Clifford, 1990) proves that Markov and factorization properties are equivalent, for any strictly positive distribution.

In order to estimate and compute the parameters of the graph, first, it is decomposed into its maximal cliques. Then the relevant coefficients of each clique are computed and finally, the entire graph is reformed considering the estimated coefficients. This chapter will concentrate on the pairwise Markov graph (Koller & Friedman, 2009) in which the dependence structure is represented by the second-order relationships, $f^{(2)}$. This type of models is usually termed as parsimonious models and are characterized by reduced complexity since fewer parameters are required to explain the data (Raykov & Marcoulides, 1999).

4.3.2.2 Gaussian Graphical Models

In Markov networks, all variables are assumed to be continuous. Furthermore, when the observations are continuous and have a multivariate normal distribution of p dimensions, $X \sim N(\mu, \Sigma)$, with mean $\mu \in \mathbb{R}^p$ and covariance matrix Σ, then the graphs are referred to as *Gaussian Graphical Models* (GGM; Lauritzen, 1996). In addition, the covariance between two variables i and j conditioned on all other variables is known as *partial covariance* and it is included in the inverse covariance matrix Σ^{-1}. The values of the Σ^{-1} range from -1 to 1. Moreover, the inverse covariance matrix, $\Theta = \Sigma^{-1}$ is commonly known as *precision* or *concentration matrix* (Hastie et al., 2015; Cox & Wermuth, 1993). Therefore, when the i and j are conditionally independent considering the rest variables, then the θ_{ij} component of Θ is zero.

To investigate the conditional distribution of one variable given the others (Hastie et al., 2009; Wasserman, 2004), the random normal vector X is partitioned as $X = (X_A, X_B)$ where $X_A = (X_1, \ldots, X_{p-1})$ consists of the first $p - 1$ variables and $X_B = X_p$ consist of the last variable (the target column). Likewise, the mean, μ, is divided in two parts $\mu = (\mu_A, \mu_B)$. With the intention of calculating the conditional distribution for X_B given $X_A = x_A$, first, the covariance matrix Σ can be written in block form, as:

$$\Sigma = \begin{pmatrix} \Sigma_{AA} & \sigma_{AB} \\ \sigma_{AB}^T & \sigma_{BB} \end{pmatrix} \tag{4.6}$$

Concerning that the marginal distribution of X_A is $X_A \sim N(\mu_A, \Sigma_{AA})$, the conditional distribution of X_B given X_A is:

$$\begin{aligned} X_B \mid X_A = x_A &\sim \mathcal{N}\left(\mu_{B|A}, \Sigma_{B|A}\right) \\ &= x_A \sim \mathcal{N}\left(\mu_B + \Sigma_{AA}^{-1}\sigma_{AB}(x_A - \mu_A)^T, \sigma_{BB} - \sigma_{AB}^T\, \Sigma_{AA}^{-1}\sigma_{AB}\right) \end{aligned} \tag{4.7}$$

It is well known that the best estimate of regression coefficient β can be found by multiplying the inverse of the variance of X_A (Σ_{AA}^{-1}) with the observed covariance of X_B and X_A (σ_{AB}), $\beta = \Sigma_{AA}^{-1}\sigma_{AB}$, assuming that the residuals of the predictors are independent ($E[X_A, e] = 0$) (McArdle & Kadlec, 2013; Hastie et al., 2009; Bhoj & Ahsanullah, 1994). Assuming that Θ is partitioned as Σ (Eq. 4.6) then $\Sigma\Theta = \text{I}$ is called as identity matrix in which the expanding relation is resulted as:

$$\begin{pmatrix} \Sigma_{AA} & \sigma_{AB} \\ \sigma_{AB}^T & \sigma_{BB} \end{pmatrix} \begin{pmatrix} \Theta_{AA} & \theta_{AB} \\ \theta_{AB}^T & \theta_{BB} \end{pmatrix} = \begin{pmatrix} I & 0 \\ 0^T & 1 \end{pmatrix} \tag{4.8}$$

which gives $\Sigma_{AA}\theta_{AB} + \sigma_{AB}\theta_{BB} = 0$. Considering that $\beta = \Sigma_{AA}^{-1}\sigma_{AB}$, it is emerged:

$$\beta = -\theta_{AB}/\theta_{BB} \tag{4.9}$$

According to the above, the multiple linear regression could determine the dependence structure. Specifically, if $\beta = 0$, then the corresponding variables of X_A are conditionally independent of the target variable of X_B, versus the rest, $\theta_{AB} = 0$. Moreover, the concentration matrix Θ includes the second-order interactions to describe the conditional distribution of each node given the rest.

Nevertheless, due to sampling variation, none of the partial correlation coefficients equal to zero. Actually, the graph obtains *spurious or false positives* connections (Costantini et al., 2015). The term spurious connection signifies a relationship among two variables which it seems to be real but is not. These connections are identified as very small (close to zero) estimates that represent very weak strengths among variables in the graph (Dikopoulou et al., 2020; Epskamp & Fried, 2017).

The next section introduces a combined method, the graphical lasso, based on global likelihood function of the graphical model and the ℓ_1-regularization which constrains or regularizes the coefficient estimates. Moreover, it selects the best model (Yuan & Lin, 2007), a problem which is known as *covariance selection*.

4.3.2.3 Graphical Lasso

The graphical lasso (Friedman et al., 2008) is an for learning the structure in an undirected Gaussian graphical model. It uses a statistical regularization technique originating in the field of machine learning, the ℓ_1-regularization, to control the spurious connections in the precision matrix. In simple words, when a Gaussian graphical model is estimated, the '*least absolute shrinkage and selection operator*' (lasso; Tibshirani, 1996) shrinks the small partial correlation coefficients exactly to zero. Consequently, spurious edges can be omitted from the graph and a sparser network structure is obtained which can explain the covariance between variables in the data.

The graphical lasso is a regularization framework for estimating the covariance matrix Σ under the assumption that precision matrix Θ is sparse (Meinshausen & Bühlmann, 2006; Yuan & Lin, 2007). The graphical lasso problem maximizes the penalized (ℓ_1-regularized) log-likelihood:

$$\underset{\Theta \succ 0}{maximize}\, f(\Theta) := \log\det(\Theta) - \mathrm{tr}(S\Theta) - \lambda\Theta_1 \tag{4.10}$$

where the derivative log det (Θ) equals to $\Theta^{-1} = W$ (Boyd & Vandenberghe, 2004), S denotes the sample covariance matrix, λ (lambda) is a nonnegative tuning parameter controlling the amount of ℓ_1 shrinkage and $||\Theta||_1$ is the ℓ_1 norm (the sum of the absolute values of the elements of Σ^{-1}). Expression (4.10) is the Gaussian log-likelihood of the data, partially maximized with respect to the mean parameter μ. The negative of this penalized likelihood is a convex function of Θ.

The gradient equation for maximizing the log-likelihood (4.10) can be written as:

$$\Theta^{-1} - S - \lambda\Gamma = 0 \tag{4.11}$$

where Γ is a symmetric matrix ($p \times p$) of component-wise signs of Θ which the diagonal elements are zero:

$$\begin{aligned} \gamma_{ij} &= sign(\theta_{ij}) \quad if\ \theta_{ij} \neq 0 \\ \gamma_{ij} &\in [-1, 1] \quad if\ \theta_{ij} = 0 \end{aligned} \tag{4.12}$$

In order to solve the Eq. (4.10), the *glasso* algorithm uses a block-coordinate method to partition Θ, S, Γ and W as shown in (Eq. 4.6). All these matrices are divided into two parts, the first part includes the first $p-1$ rows and columns and the second part obtain the pth row and column. Moreover, the algorithm solves for a row/column of (Eq. 4.11) at a time, holding the rest fixed. For simplicity, we focus on the last row and column; therefore, the upper right block of the equation in (4.11) leads to (4.13).

$$w_{12} - s_{12} - \lambda\gamma_{12} = 0 \tag{4.13}$$

Since, $\Sigma\Theta = I$, (in Eq. 4.8), the corresponding relation of the partitioned W and its inverse Θ is $W\Theta = I$. Additionally, using properties of inverses of block-partitioned matrices, $W = \Theta^{-1}$ can be written as in Eq. (4.14). The equivalent equation of w_{12} recovered from Eq. (4.14) is presented in Eq. (4.15).

$$\begin{pmatrix} W_{11} & w_{12} \\ w_{21} & w_{22} \end{pmatrix} = \begin{pmatrix} \left(\Theta_{11} - \frac{\theta_{12}\theta_{21}}{\theta_{22}}\right)^{-1} & -W_{11}\frac{\theta_{12}}{\theta_{22}} \\ . & \frac{1}{\theta_{22}} - \frac{\theta_{21}W_{11}\theta_{12}}{\theta_{22}^2} \end{pmatrix} \tag{4.14}$$

$$w_{12} = -W_{11}\frac{\theta_{12}}{\theta_{22}} = W_{11}\beta \tag{4.15}$$

The Eqs. (4.9) and (4.15) are substituting into Eq. (4.13) which gives $W_{11}\beta - s_{12} - \lambda\gamma_{12} = 0$. Moreover, two more mathematical expressions are retrieved, $\beta = w_{12}/W_{11}$ and $\theta_{12} = -\theta_{22}w_{12}W_{11}^{-1}$. Since $\theta_{22} > 0$, it is proven that β and θ_{12} have opposite signs; $sign(\theta_{12}) = -sign\left(w_{12}W_{11}^{-1}\right) = -sign(\beta)$, which formalize the equation below:

$$W_{11}\beta - s_{12} + \lambda sign(\beta) = 0 \tag{4.16}$$

Assuming that $W_{11} \succ 0$ is fixed, the ℓ_1-regularized quadratic program of the (Eq. 4.16) is:

$$\underset{\beta \in \Re^{p-1}}{minimize}\left\{\frac{1}{2}\beta' W_{11}\beta + \beta' s_{12} + \lambda\|\beta\|_1\right\} \tag{4.17}$$

The above equation is similar to a lasso regression problem which identifies the relationship between the last variable (the dependent variable) given the rest (independent variables or predictors). The only exception is the usage of estimate W_{11} of the upper block of W instead of the cross-product matrix S_{11} (Friedman et al., 2008). This problem can recover the estimate $\widehat{w}_{12}$ solving the equation of $\widehat{\beta}$ from Eq. (4.15). Afterwards, the lower-right element of (4.14) retrieves the $\widehat{\theta}_{12}$ measurement from $\widehat{\beta}$ and $\widehat{\theta}_{22}$.

$$\frac{1}{\widehat{\theta}_{22}} = w_{22} - \widehat{\beta}'\widehat{w}_{12} \tag{4.18}$$

It is noteworthy to mention that solving β and updating w_{12}, the *glasso* algorithm repeats this procedure for all variables until convergence. An iterative algorithm is converged or reached the equilibrium point when the output approaches as possible to a specific value as the iterations proceed. Mainly, an algorithm converges to the desired solution from any given starting point according to a residual (or error) value which is known as globally convergent (Jensen et al., 1991). The final algorithmic rule is the estimation of θ_{12} and θ_{22} (Mazumder & Hastie, 2012). Moreover, it should be emphasized that the transform solution from Eq. (4.11) for all i is $w_{ii} = s_{ii} + \lambda$, since $\theta_{ii} > 0$, and $\Gamma_{ii} = 0$. The *glasso* algorithm proposed by Friedman et al. (2008) relying on the research of Banerjee et al. (2008) is outlined in Algorithm 4.1.

Algorithm 4.1 *Graphical lasso via inverse covariance structure.*

1. Initialize $W = S + \lambda I$. The diagonal of W remains unchanged in what follows.
2. Partition the matrix W into two parts. Part 1 includes all columns except p^{th} row and p^{th} column and part 2 consists of the pth row and column. Perform the following steps till convergence:
   a. Rearrange the rows/columns so that the target column (part 2) is last.
   b. Solve the lasso problem (Eq. 4.17), using as initial value the solution from the previous round for this column.
   c. Update the row/column (off-diagonal) of the covariance using $\widehat{w}_{12}$ (Eq. 4.15).
   d. Save $\widehat{\beta}$ for this column in the matrix B.
3. Finally, for every row/column, calculate the diagonal entries $\widehat{\theta}_{jj}$ using Eq. (4.18), and transform the B matrix to Θ.

As described above, a tuning parameter or penalty parameter, λ, is used to control the strength of the penalty term in the lasso regression. In other words, it regulates the sparsity of the network. Generally, different values of λ estimate different network structures (Zhao & Yu, 2006) indicating that lasso estimates a group of networks ranging from a fully connected network (λ_{min}) to an empty network (λ_{max}). More precisely, as the tuning parameter increases, many edges are omitted from the network. On the contrary, as the tuning parameter decreases, fewer edges are removed. Subsequently, an optimal network with many true connections and few spurious connections can be obtained through model selection (Drton & Perlman, 2004). This selection can be performed by minimizing the Extended Bayesian Information Criterion (EBIC) to optimize the fit of the model to the data:

$$EBIC = -2L + Elog(n) + 4\gamma Elog(p) \tag{4.19}$$

As shown in Eq. (4.19), the EBIC adds an extra penalty, the hyperparameter γ (gamma) to control the model complexity (Foygel & Drton, 2010; Chen & Chen, 2008), L indicates the log-likelihood, n the sample size, E the number of non-zero edges and p the number of nodes. Moreover, the hyperparameter γ is set manually, ranging from 0 to 0.5 (Foygel & Drton, 2010). It is worth noting that the hyperparameter γ must not be confused with the tuning parameter λ. Setting γ equal to zero, the EBIC is reducing to the standard BIC, which retrieves a sparser graph compared to a partial correlation network. To conclude, the combination of lasso regularization with EBIC model selection has been revealed to identify the true network structure, particularly when a sparser network is created (Epskamp & Fried, 2017; Epskamp, 2016; Foygel & Drton, 2010).

This study applies the '*qgraph*' R package (Epskamp et al., 2012) which integrates the *lasso* algorithm with the EBIC model selection. R is a powerful language and environment for statistical and graphical techniques. The *qgraph* package utilizes the part of the glasso algorithm (Friedman et al., 2008) from the '*glasso*' package (Friedman et al., 2014). Generally, it estimates 100 different models (which is the default value) corresponding to 100 different λ values and the model that minimizes the EBIC is the one that fits better to the data (Fig. 4.2). On a final note of

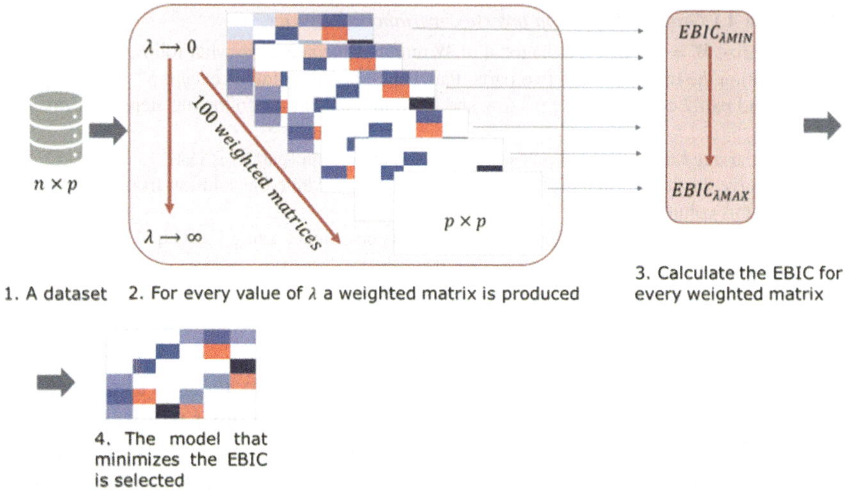

Fig. 4.2 The algorithm of the regularized graphical model

this research, the hyperparameter γ is set equal to 0.5 as Foygel and Drton (2010) suggest.

4.3.2.4 Graphical Lasso Handling Ordinal Data

As it is mentioned above, the estimation of any partial correlation network requires multivariate normal distributions. This study handles data characterized by ordinal scales, which occur difficulties in measurement since they cannot be added or multiplied. The question that reveals is how is it possible to estimate partial networks from ordinal data. The answer was given a long time ago by Karl Pearson (1900). He proposed a method to present the data for pairs of ordinal variables in the form of contingency tables. Each value of an ordinal variable represents either a row or a column and frequencies are presented in the corresponding cells. Later, these correlations were termed as polychoric correlations (Olsson, 1979).

The principal concept of the polychoric correlation coefficient is a measure of association for ordinal variables. Specifically, the ordinal variables are considered as discretized random variables which rely on an assumption of an underlying joint normal distribution (Muthén, 1984; Pearson, 1900). Pearson visualized the contingency table with the bell-shaped bivariate normal density function standing on top of it. In simple words, the vertical lines include the cells of the contingency table which classify the parts of the bivariate normal distribution according to the categories of the ordinal data. When the thresholds (the vertical lines) are estimated then the correlations between latent variables can be estimated pairwise. Figure 4.3 illustrates an example of the threshold function as presented in the work of Epskamp & Fried (2017).

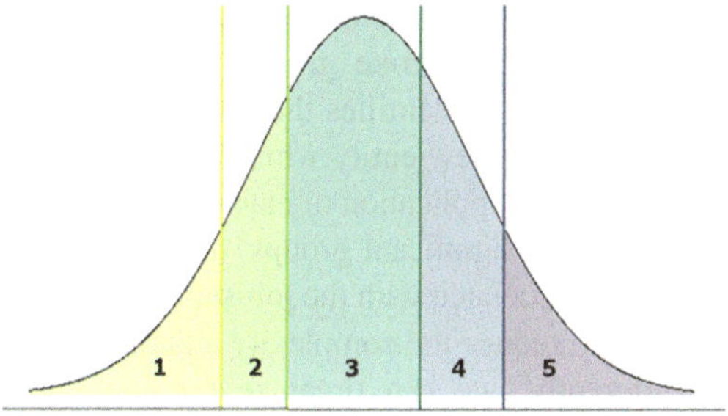

Fig. 4.3 Visualization of the domain of the standard normal density function discretized by the vertical lines which represent the thresholds

In this figure, three thresholds are estimated to cut the normal distribution in four segments. The normal distribution corresponds to the latent item score and rectangles represent the thresholds. For the sake of completeness, the variables are measured under a five-point Likert scale ranging from 1 to 5. Therefore, a participant with a latent score lower than the first threshold would be equal to one, a participant with a latent score between the first and the second threshold would be rated as two, and so on. The major benefit of the qgraph package is the automatic identification of the type of data. Consequently, when the variables are ordinal, the package estimates the polychoric correlation matrix which is helpful to measure the covariance matrix (Epskamp et al., 2012).

4.4 Modelling on Business Network Structures

In this section, we present the resulted network structures after implementing the distance-based method proposed by Schneider et al. and Graphical Models, respectively. It is mentioned that the initial matrices concerning the job satisfaction problem are the same for both methods. For example, the dimensions of the initial dataset regarding the automotive sector (S_1) are 10×1043, where ten determine the number of variables/concepts (V_1–V_{10}; Table 2.1) and 1043 the number of individuals ($N = 1043$; Table A.1).

The R-package *qgraph* (Epskamp et al., 2012) has been used to visualize the graphs of the two methods. Nodes that are depicted closer together are strongly related. Hence, wider and more saturated edges determine stronger relations; while, positive and negative relations are represented by blue and red connections, respectively. Thus, the *qgraph* package includes the Fruchterman-Reingold's algorithm (Fruchterman & Reingold, 1991). This efficient algorithm locates the nodes of the network according to the link weights to clarify the connectivity of the network. Specifically, strongly connected nodes are placed closer and weakly related nodes are further away from each other.

As already mentioned, not all complex problems could be explained as a causal view of latent factors. Some problems are more appropriate to be represented as networks, which model the most important and direct relations between the observed variables. For this reason, it is consequential to identify if and how the variables are partitioned into discrete groups using the clustering analysis. Clustering is the procedure which identifies the important groups in the data which the information is reduced to a new entity which is called cluster (Kotu & Deshpande, 2019). The most common application of clustering is to explore the data in order to detect all the potential and significant groups. Consequently, we would like to investigate if the variables associated with the job-satisfaction problem present any similar properties/patterns to reduce the complexity and facilitate the interpretation of our data. To put it differently, we can organize the job-attractiveness variables into clusters to determine the internal structure of the data.

A well-known dimension reduction method is the Exploratory Factor Analysis (EFA). It belongs to a family of multivariate analysis methods in which p measurable variables (indicators) are represented from r unobserved latent variables (common factors or hidden causes) where $r < < p$ (Mair, 2018; Thompson, 2007; Thurstone, 1931). The goal of factor analysis is to model the interrelationships between items with fewer (latent) variables by sharing the amount of underlying variance (causal impact) among a set of items and the latent variable. Preliminary, it is necessary to specify the appropriate number of factors (groups) to perform factor analysis. In general, the factors which have a high eigenvalue (equally or higher than 1) could be retained, while those with a low eigenvalue could be eliminated. A reason for this is that an eigenvalue of one explains the variance of a single variable (Kaiser, 1960; Braeken & van Assen, 2017). Consequently, factors that explain at least the same amount of variance as a single variable are beneficial. This decision is often motivated or supported by some statistical measurements and routines in order to indicate the optimal number of factors. In this study, we apply the 'phycho' R package which runs nine different routines and returns the possible number of groups with the highest consensus.

For example, in the Automotive sector (Fig. 4.4), six out of nine methods obtain the eigenvalue higher than 1 ($Eigen = 5.41$) and the cumulative proportion of explained variance over 0.5 ($Cumulative.\ var_{S1} = 0.541$) with only 1 factor (cluster or group). This indicates that the variables of the dataset are grouping all together. The results of the rest sectors of this study indicate as well that the majority of the methods[1] (six out of nine; 66.67%) determine that the cumulative variance ($0.526 \leq Cumulative.\ var \leq 0.596$) of the variables could be as well explained with only one cluster. Accordingly, it is inferred that our data cannot be represented using any dimensional reduction method since the variables do not show any clustering characteristics. Nevertheless, for the sake of convenience, the explained variance can be improved by increasing the number of clusters to maintain the clusters in equal sizes or densities (Akerkar & Lingras, 2007). Therefore, we would like to investigate the appropriate maximum number of clusters that each of

[1] Optimal Coordinates, Acceleration Factor, Parallel Analysis, Eigenvalues (Kaiser Criterion), Velicer MAP, VSS Complexity 1.

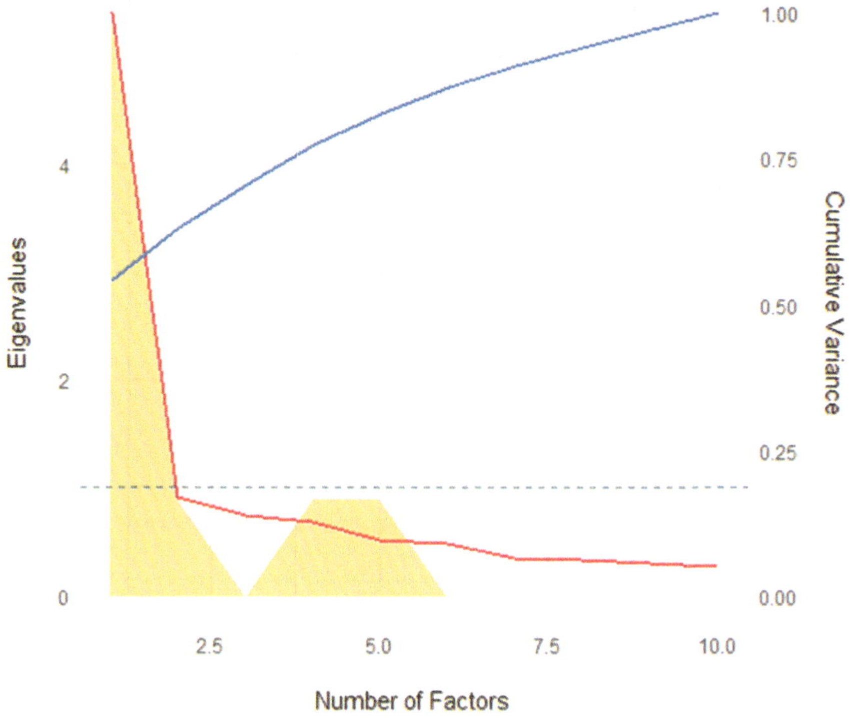

Fig. 4.4 The plot shows the number of methods (in yellow), the eigenvalues (red line) and the cumulative proportion of explained variance (blue line) of the Automotive sector

them will include two or more items (Johnson, 1967). It is notable to refer that the percentage of variance is explained for different cluster solutions.

One of the simplest unsupervised learning algorithms taking as input the number of clusters is the *k*-means clustering. The objective of *k*-means algorithm is to find the centroids, the centres of *k* clusters. Each centroid is defined by the mean of all data objects in the cluster (Lloyd, 1982). Visually, the *k*-means divide the data space into *k* groups in *n*-dimensional space, where *n* indicates the number of variables in a given dataset. Finally, the *k*-means clustering algorithm converges and the data points are divided into *k* clusters.

Precisely, the data points are assigned to their closest centroid according to Euclidean distance between the object and the cluster mean. For this reason, two more R packages are applied to the dataset, the '*cluster*' (Maechler et al., 2019) and the '*factoextra*' (Kassambara & Mundt, 2017). The first one estimates the clusters of the observed variables using the "kmeans" function and the other visualizes the outcome (Figs. 4.5 and 4.6). Because the number of clusters (*k*) is set before the execution of the algorithm, it is often considered to use several different values of *k* and investigate the differences in the results. Moreover, for this study, the "the *Hartigan and Wong* (Hartigan & Wong, 1979) algorithm was applied to identify the

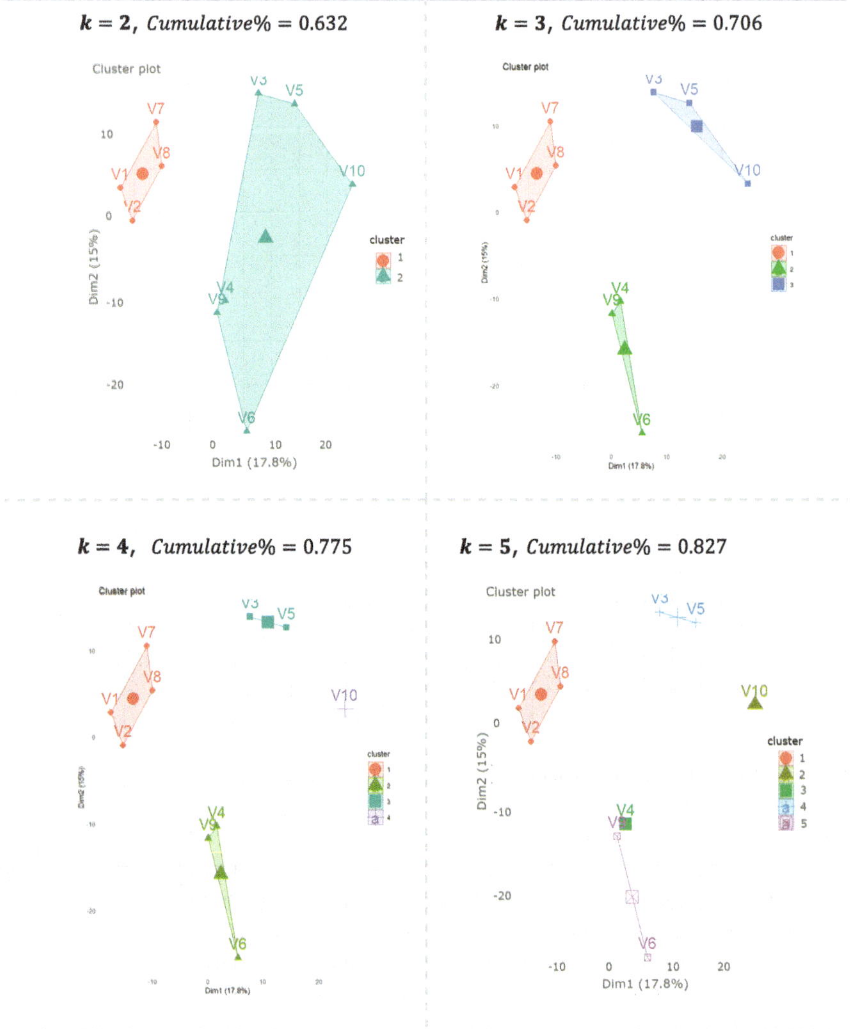

Fig. 4.5 The job-satisfaction variables of the Automotive sector (S_1) applying the *k*-means clustering (*k*) of 2, 3, 4 and 5 clusters

centroids. In particular, it seeks for the partition of data space with a locally optimal within-cluster sum of squares of errors and it updates the centroids any time a point is moved.

Figure 4.5 depicts how the variables in the Automotive sector are grouping from two to five clusters and the correspondent cumulative variance. Clearly, when the number of clusters is four, item V_{10} tends to cluster in a separate group even though it is located closer to the cluster that includes V_9 and V_6.

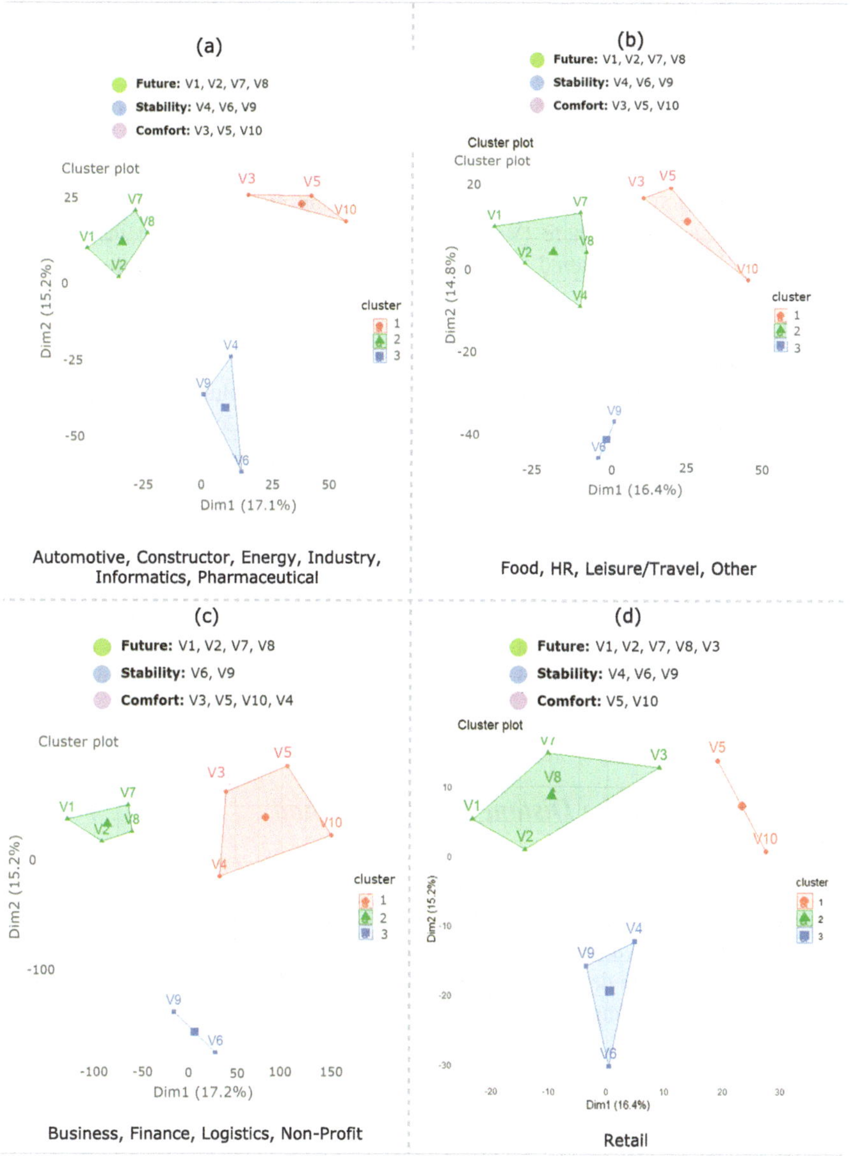

Fig. 4.6 Visualization of three clusters of ten observed job-satisfaction variables using *k*-means clustering method. Each Sector belongs to one of the four groups (**a**–**d**)

Similarly, in five clusters, items V_{10} and V_4 are separated to different clusters. This violates the assumption of a cluster which consists of two or more items. On the contrary, when the items are grouped in two clusters, the cumulative variance is 0.632, but the items in the blue cluster are depicted far from each other. For this reason, the considerable and maximum number of groups is three. In this case, the cumulative proportion of explained variance is increased to 0.706. As a clarification, the groups are setting equal to three just for the visual purpose, as the data-driven routines (Makowski, 2018) indicate that the variables are grouping only in one cluster.

To examine which variables concerning the job-satisfaction are grouped, Fig. 4.6 illustrates three clusters determined by the *k*-means method. It is observed that Pleasant working atmosphere (V_3) and Long-term security (V_4) are inconsistent. Particularly, V_3 is clustered except with Comfort, also with Future cluster (Fig. 4.6d); while V_4 is grouped apart from Stability as well as with Future (Fig. 4.6b) and Comfort (Fig. 4.6c) clusters. For example, in the Automotive sector, the first cluster (green color) consists of four variables (V_1, V_2, V_7, V_8) and it is associated with the Future substance. The second cluster (blue color) is related to Stability (V_4, V_6, V_9) and the last one (red color), with Comfort (V_3, V_5, V_{10}). Therefore, the Automotive sector belongs to the first group (Fig. 4.6a) Moreover, Fig. 4.6 shows how the 15 sectors are classified into four groups according to the clusters obtained from *k*-means clustering method. For instance, the third group, (Future: V_1, V_2, V_7, V_8; Stability: V_6, V_9; Comfort: V_3, V_5, V_{10}, V_4) includes the following four working sectors: Business, Finance, Logistics, Non-Profit (Fig. 4.6c). The fourth group contains only one, the Retail sector, as it is presented in Fig. 4.6d. To recall, the variables are not providing significant differences from cluster to cluster. The reason that is used is to assist in the interpretation of the results.

4.4.1 Results of the Distance-Based Method

The distance-based algorithm (Schneider et al. 1998) was applied to 15 real datasets corresponding to 15 sectors as described in Chap. 2 in order to define relations among the observed variables associated with the job attractiveness in Belgium. Due to the large size of the datasets, we have implemented the distance-based method in the R programming language to facilitate the calculations of thousands of participants of this method.

After the determination of the initial matrix (Table 4.3), the fuzzified matrix is constructed based on Eq. (4.1; Table 4.4). Threshold values are set to correct the possible deviation of ±20% between the variables of the interviewed employees. For example, the evaluation of the fifth individual for the third variable (V_3: Pleasant working environment) is the element O_{35} (Table 4.4). The transformed fuzzy value, after the threshold adjustment is equal to 1, since $X_3(O_{3,5}) = \frac{5-1}{4-1} = 1.333 = 1$. The fuzzy value 1 indicates a high degree of membership of the V_3 variable regarding the opinion of the fifth participant. The required calculations were accomplished, applying the Eqs. (4.2)–(4.4) to obtain the symmetric SIRM matrix [10 × 10]. Each element S_{ij} express the strength of the relationship between two concepts. For example, the $S_{28} = 0.836$ denotes that the concept V_2 (Prospects/career

Table 4.3 The Initial Matrix of the Automotive Sector that is used to calculate the network applying the distance-based method

Variables	P_1	P_2	P_3	P_4	P_5	P_6	P_7	P_8	...	P_{1043}
	Individuals									
V_1	4	1	1	2	5	4	4	5	...	1
V_2	1	1	1	2	5	4	4	4	...	1
V_3	2	4	3	1	5	1	1	1	...	1
V_4	1	1	5	5	5	4	1	5	...	5
V_5	4	4	5	3	3	1	1	2	...	5
V_6	5	4	2	5	5	1	4	5	...	5
V_7	1	1	3	4	5	1	1	4	...	5
V_8	2	1	2	2	2	4	4	3	...	1
V_9	4	4	1	1	3	1	1	1	...	1
V_{10}	3	3	2	4	1	1	1	4	...	1

Table 4.4 The Fuzzified Influence Matrix (FIM) of the distance-based method of the Automotive Sector

Variables	P_1	P_2	P_3	P_4	P_5	P_6	P_7	P_8	...	P_{1043}
	Individuals									
V_1	1.000	0.000	0.000	0.333	1.000	1.000	1.000	1.000	...	0.000
V_2	0.000	0.000	0.000	0.333	1.000	1.000	1.000	1.000	...	0.000
V_3	0.333	1.000	0.667	0.000	1.000	0.000	0.000	0.000	...	0.000
V_4	0.000	0.000	1.000	1.000	1.000	1.000	0.000	1.000	...	1.000
V_5	1.000	1.000	1.000	0.667	0.667	0.000	0.000	0.333	...	1.000
V_6	1.000	1.000	0.333	1.000	1.000	0.000	1.000	1.000	...	1.000
V_7	0.000	0.000	0.667	1.000	1.000	0.000	0.000	1.000	...	1.000
V_8	0.333	0.000	0.333	0.333	0.333	1.000	1.000	0.667	...	0.000
V_9	1.000	1.000	0.000	0.000	0.667	0.000	0.000	0.000	...	0.000
V_{10}	0.667	0.667	0.333	1.000	0.000	0.000	0.000	1.000	...	0.000

opportunities) directly influence the concept V_8 (Offers good quality of training). The weight matrix of the SIRM is presented in Table 4.5 with the correspondent visualization graph (Fig. 4.7).

The density of the graphs is associated with their interpretability; generally, the transparency is increasing as the number of connections between vertices (nodes) of the graph is decreasing (Scott, 2000). For this reason, the structures of the models which are generated from real data are compared by measuring the density of each map. The network *density* is estimated by dividing the number of actual edges in the network, over the maximum number of the edges between all pairs of nodes ($\frac{n(n-1)}{2}$). Generally, the density of a complete graph is equal to one ($\delta = 1$), in which all nodes

Table 4.5 The SIRM matrix of ten variables correlated to a job satisfaction problem of the Automotive sector (S_1)

Automotive

	V1: Competitive salary package	V2: Prospects/career opportunities	V3: Pleasant working environment	V4: Offers long-term job security	V5: Good balance (private life & work)	V6: Financially sound	V7: Offers interesting jobs	V8: Offers good quality of training	V9: Strong management	V10: Deliberately handles the environment and society
V_1	-	0.849	0.785	0.789	0.777	0.778	0.829	0.831	0.789	0.742
V_2	0.849	-	0.795	0.821	0.770	0.784	0.829	0.836	0.803	0.750
V_3	0.785	0.795	-	0.785	0.826	0.771	0.826	0.802	0.807	0.804
V_4	0.789	0.821	0.785	-	0.781	0.812	0.781	0.793	0.779	0.771
V_5	0.777	0.770	0.826	0.781	-	0.762	0.788	0.788	0.766	0.792
V_6	0.778	0.784	0.771	0.812	0.762	-	0.763	0.773	0.828	0.761
V_7	0.829	0.829	0.826	0.781	0.788	0.763	-	0.841	0.799	0.769
V_8	0.831	0.836	0.802	0.793	0.788	0.773	0.841	-	0.810	0.781
V_9	0.789	0.803	0.807	0.779	0.766	0.828	0.799	0.810	-	0.779
V_{10}	0.742	0.750	0.804	0.771	0.792	0.761	0.769	0.781	0.779	-

are connected. Such dense graphs are very unusual, even in very small networks (Scott, 2000) and cannot interpret efficiently the real problem.

The derived job-satisfaction network of the Automotive sector is characterized as a complete graph since each node is linked to all other nodes of the graph. Consequently, the link density δ of this dense network is one which is the maximal value. Moreover, only positive and implausibly high connections (the majority of weights is higher than 0.8) are included. All these conditions affect the interpretability of this network which make the interpretation of the structures difficult to understand (Scott, 2000).

Applying the distance-based method to our dataset, all job-satisfaction networks corresponding to 15 working sectors are specified as complete graphs ($\delta = 1$) indicating that all variables are connected to each other. In addition, the weights of 15 sectors are high fluctuating between 0.724 and 0.864 (0.8 ± 0.27, $n = 675$). Moreover, in Fig. 4.8, the nodes are not in agreement according to k-means clustering method (Figs. 4.5 and 4.6). Under these circumstances, it is concluded that the distance-based method cannot provide interpretable networks when the data sample is high (over 50 observations).

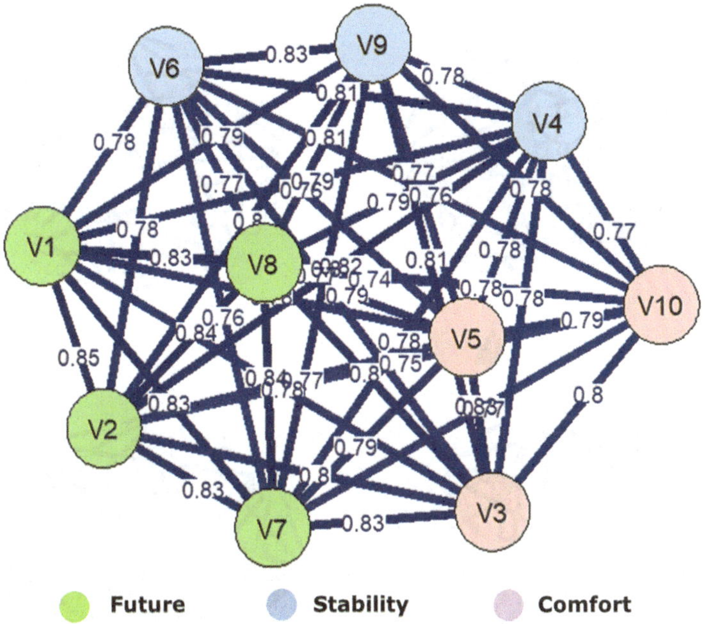

Fig. 4.7 Graph visualization of distance-based method applied in Automotive dataset. The derived network is considered as a complete graph since $\delta = 1$. The nodes are colored according to the outcome of the k-means clustering method (Fig. 4.5)

4.4.2 Results of Graphical Lasso

Employees from 15 different sectors in Belgium rated the importance of ten job-satisfaction factors, measured on a five-point scale. A detailed description of the dataset can be found in Chap. 2. The graphical lasso method in combination with EBIC model selection (Foygel & Drton, 2010) are performed to estimate 15 Gaussian graphical models from 56,246 observations. This procedure requires a measurement of the variance-covariance matrix and returns a parsimonious network of partial correlation coefficients (Epskamp & Fried, 2017).

As Epskamp and Fried (2017) recommend, when data are ordinal, the polychoric correlation matrix must be computed. This process could be performed automatically using the *cor_auto* function from qgraph (Epskamp et al., 2012) package which detects the type of the variables and performs the correspondent correlation. To recall, the number of the expected estimated networks depended on the value of the tuning parameter, $=100$. The λ regulates the amount of the penalization from completely empty to completely connected network. Next, the EBICglasso function from qgraph package selects the network structure that minimizes the EBIC. In short, EBIC, adds an extra penalty, the hyperparameter (or tuning parameter) γ to control the model's complexity. This tuning parameter ($0 \leq \gamma \leq 1$) controls the sparsity of the graphs (Foygel & Drton, 2010). As the number of γ increases, the sparser the

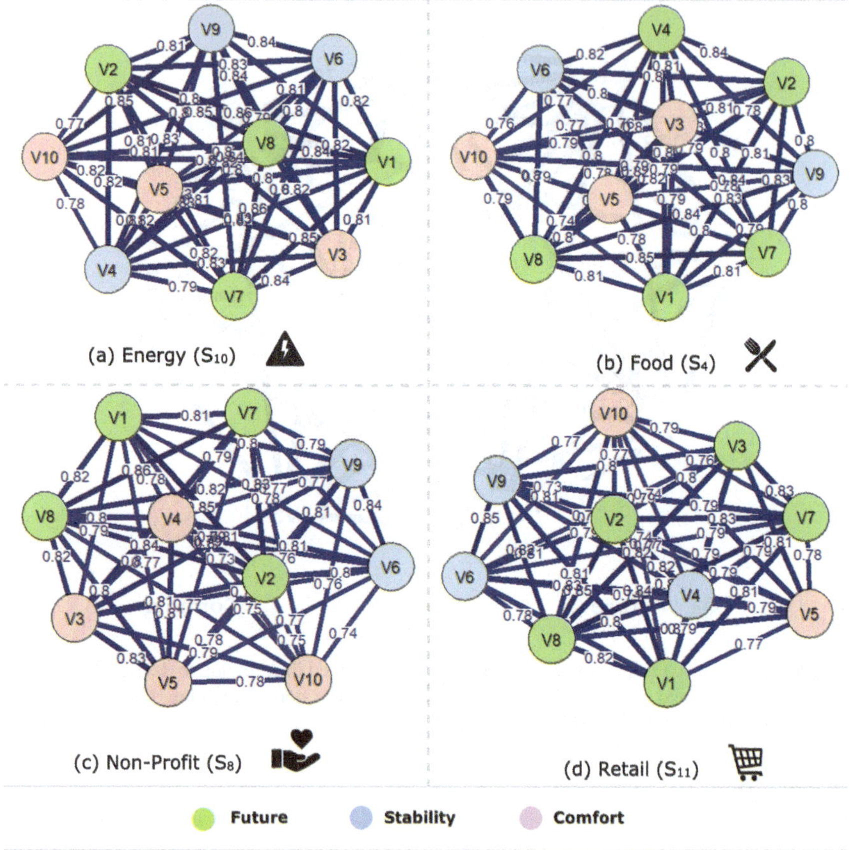

Fig. 4.8 Visualizations of complete structures ($\delta = 1$) of four sectors using the distance-based method. The nodes of these networks are colored according to the outcome of the *k*-means clustering method (Figs. 4.5 and 4.6)

graph will be. However, Foygel and Drton have proven that if γ fluctuating between 0 and 0.25 then the false positives will be decreased, without increasing the false negatives. Generally, the aim of the model selection is to identify the conditional independent relationships (the zero edges). Nevertheless, some very weak edges are estimated in the model. Therefore, the argument threshold transforms the weak edges that are not significantly different to zero to be equal to zero.

The qgraph plots the sparser graphical models of ten nodes (Fig. 4.9). Each node in the network represents a job-satisfaction factor (variable) and each edge depicts the partial correlation between the factors controlling for all other connections in the network. Figure 4.9 illustrates the partial correlations of the Automotive sector of Table 4.6. This sector estimates only 28 non-zero edges out of 45 possible edges. In particular, a strong connection ($w_{12} = 0.34$) is emerged among nodes: Competitive salary package (V_1) and Career opportunities (V_2). This edge suggests that the

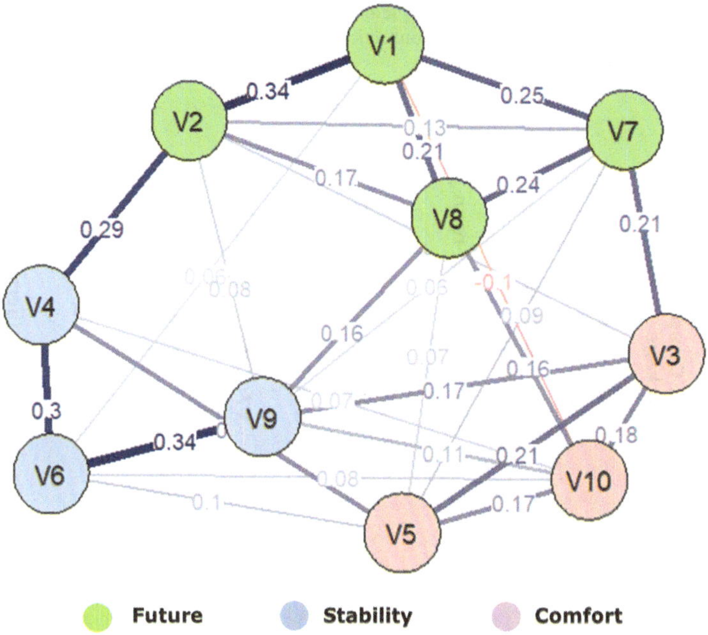

Fig. 4.9 Graph visualization of graphical lasso method applied in Automotive dataset. Blue and red edges represent positive and negative associations, respectively. Wider and more saturated edges indicate higher strengths. The colors of the nodes are depending on the *k*-means results (Fig. 4.5)

Competitive salary is a good predictor of Career opportunities and vice versa. Other strong relations are observed between Financially sound (V_6) and Strong management (V_9) with $w_{69} = 0.336$ and Long-term job security (V_4) and Financially sound (V_6) with $w_{46} = 0.3$. Medium strength connections ($0.2 \leq w < 0.3$), weak ($0.1 \leq w < 0.2$) and very weak ($0.0 < w < 0.1$) are also estimated; while a negative weak association ($w_{46} = -0.1$) is distinguished between Competitive salary package (V_1) and Deliberately handles the environment and society (V_{10}). This negative edge signifies that when V_1 increases then V_{10} decreases and vice versa. Having investigated the estimated edges, it is also reasonable to focus on the 17 absent connections ($w_{ij} = 0$). The lack of connectivity between two variables denotes that these two variables are statistically independent when conditioning on all other variables in the network. Consequently, it does not emerge any causal relationship between these variables.

Accordingly, the interpretation considering the strength of edges could be straightforward. For example, as mentioned above, the Competitive salary package (V_1) is a good predictor ($w_{12} = 0.34$) to Career opportunities (V_2). Furthermore, it is a moderate predictor with $w_{17} = 0.25$ to Interesting job (V_7) and with $w_{18} = 0.21$ to Good quality of training (V_8). Additionally, it is a very weak predictor ($w_{16} = 0.06$) to Financially sound (V_6) and a weak negative predictor ($w_{1,\ 10} = -0.1$) to

Table 4.6 The weight adjacency matrix of ten variables using the graphical lasso in the Automotive sector (S_1). Moreover, 17 out of 45 connections are distinguished as spurious (the weights are estimated equally to zero)

Automotive

	V1: Competitive salary package	V2: Prospects/career opportunities	V3: Pleasant working environment	V4: Offers long-term job security	V5: Good balance (private life & work)	V6: Financially sound	V7: Offers interesting jobs	V8: Offers good quality of training	V9: Strong management	V10: Deliberately handles the environment and society
V_1	-	0.340	0	0	0	0.059	0.248	0.215	0	-0.097
V_2	0.340	-	0.082	0.294	0	0	0.129	0.170	0.083	0
V_3	0	0.082	-	0	0.211	0	0.214	0	0.168	0.180
V_4	0	0.294	0	-	0.172	0.299	0	0	0	0.071
V_5	0	0	0.211	0.172	-	0.1	0.091	0.070	0	0.173
V_6	0.059	0	0	0.299	0.1	-	0	0	0.336	0.083
V_7	0.248	0.129	0.214	0	0.091	0	-	0.24	0.063	0
V_8	0.215	0.170	0	0	0.070	0	0.24	-	0.163	0.162
V_9	0	0.083	0.168	0	0	0.336	0.063	0.163	-	0.114
V_{10}	-0.097	0	0.180	0.071	0.173	0.083	0	0.162	0.114	-

Deliberately handles the environment and society (V_{10}). Nevertheless, V_1 is conditionally independent (absent of edges) to four variables, V_3, V_4, V_5 and V_9, after conditioning all other variables in the network.

Figure 4.10 depicts four partial correlation networks which are estimated with the graphical lasso using the EBIC method from each group as resulted from the *k*-means clustering method. However, in this dissertation, they are not presented only sparser networks structures, but also intermediate and denser structures. According to Epskamp and Fried (2017), it is inequitable to compare the derived networks regarding their density/connectivity because the EBIC is determined by the differences in sample size. Specifically, the larger is the sample, the less parsimonious the model is (Table 4.7). Notwithstanding this is a limitation, it worth refers to some notable similarities as well as considerable differences.

Moreover, it is more convenient to compare the clusters identifying *how good predictor* is a particular cluster according to the strength of weights among the nodes in the graph. For example, in Fig. 4.10d, in the Retail sector, the *Stability* cluster is defined as the strongest cluster of the graph with mean value equal to 0.282 ($\overline{w}_{Stabilty} = 0.282$). Specifically, two out of three weights are strong ($w \geq 0.3$), V_4-V_6 ($w_{4,\,6} = 0.347$) and V_6-V_9 ($w_{6,\,9} = 0.411$). Next, it is noticed that *Future*

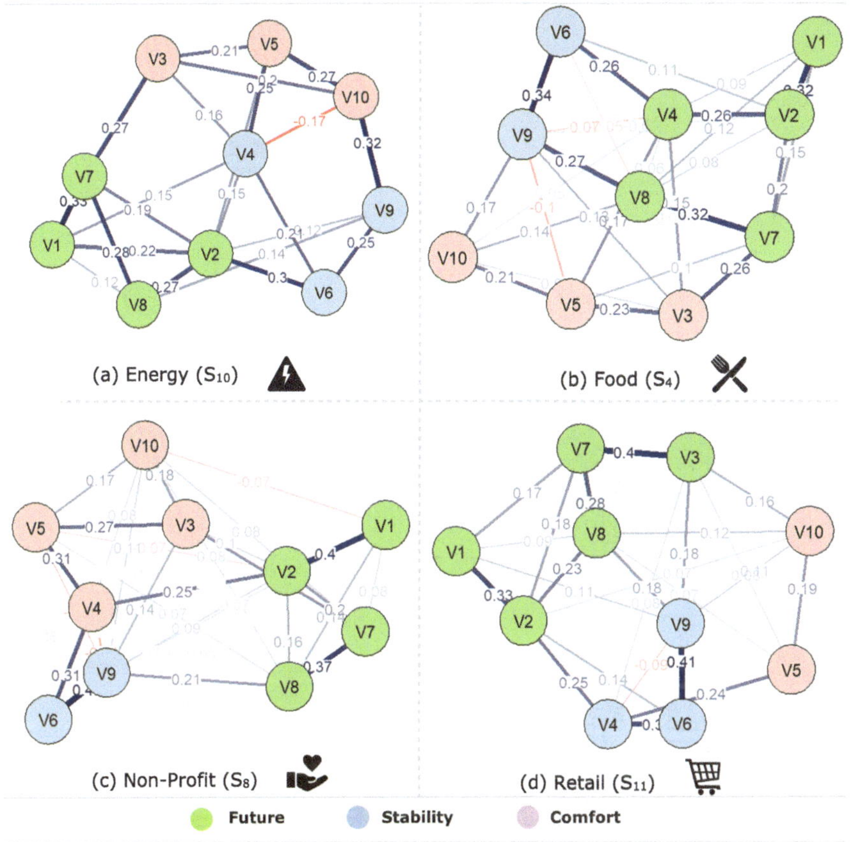

Fig. 4.10 Graph visualization of graphical lasso method applied in four sectors. The connectivity of the estimated networks ($\delta < 0.89$) indicates sparser structures comparing to complete structures ($\delta = 1$). The nodes are colored according to the outcome of the k-means clustering method (Figs. 4.5 and 4.6)

cluster consists of both high and low relations. The range of this cluster is estimated between 0.092 and 0.397. Thus, the average weights of the variables that constitute the Future cluster is $\overline{w}_{Future} = 0.236$ (lower than Stability). Therefore, Future is specified as medium predictor cluster for the Retail sector. Finally, the weight among nodes that correspond to Comfort (V_5 and V_{10}) is low ($\overline{w}_{Comfort} = 0.187$).

Hence, this cluster is characterized as a weaker predictor cluster comparing to the other two clusters. Evidence in support of the networks' similarities is summarized in Fig. 4.11. This plot highlights the associations of the weights between two variables of each sector. In this plot, every weight is colored according to the value. Particularly, there are seven categories from Positive Strong to Negative Weak which lower and upper bounds of each category determined by different colors in the palette (from blue to red color). Specifically, positive weights are displayed in blue and negative weights in red color, while zero weights are illustrated

Table 4.7 The actual number of edges and the corresponding density of each graph for every sector estimated by the distance-based and graphical lasso models

		Distance-based		Graphical Lasso				
	N. of observations	Density (δ)	N. of connections	Density	N. of connections	Positive (+) connections	Negative (−) connections	Zero (0) connections
Automotive	1043	1	45	0.622	28	27	1	17
Constructor	2301	1	45	0.778	35	34	1	10
Finance	3819	1	45	0.800	36	32	4	9
Food	1655	1	45	0.711	32	28	4	13
HR	1499	1	45	0.667	30	27	3	15
Industry	3053	1	45	0.733	33	29	4	12
Informatics	3262	1	45	0.756	34	30	4	11
Non-profit	20,378	1	45	0.889	40	35	5	5
Pharmaceutical	2363	1	45	0.644	29	28	1	16
Energy Services	172	1	45	0.489	22	21	1	23
Retail	1469	1	45	0.533	24	23	1	21
Business	3134	1	45	0.711	32	30	2	13
Transportation	2833	1	45	0.711	32	30	2	13
Travel	1598	1	45	0.600	27	27	0	18
Other	7667	1	45	0.844	38	35	3	7

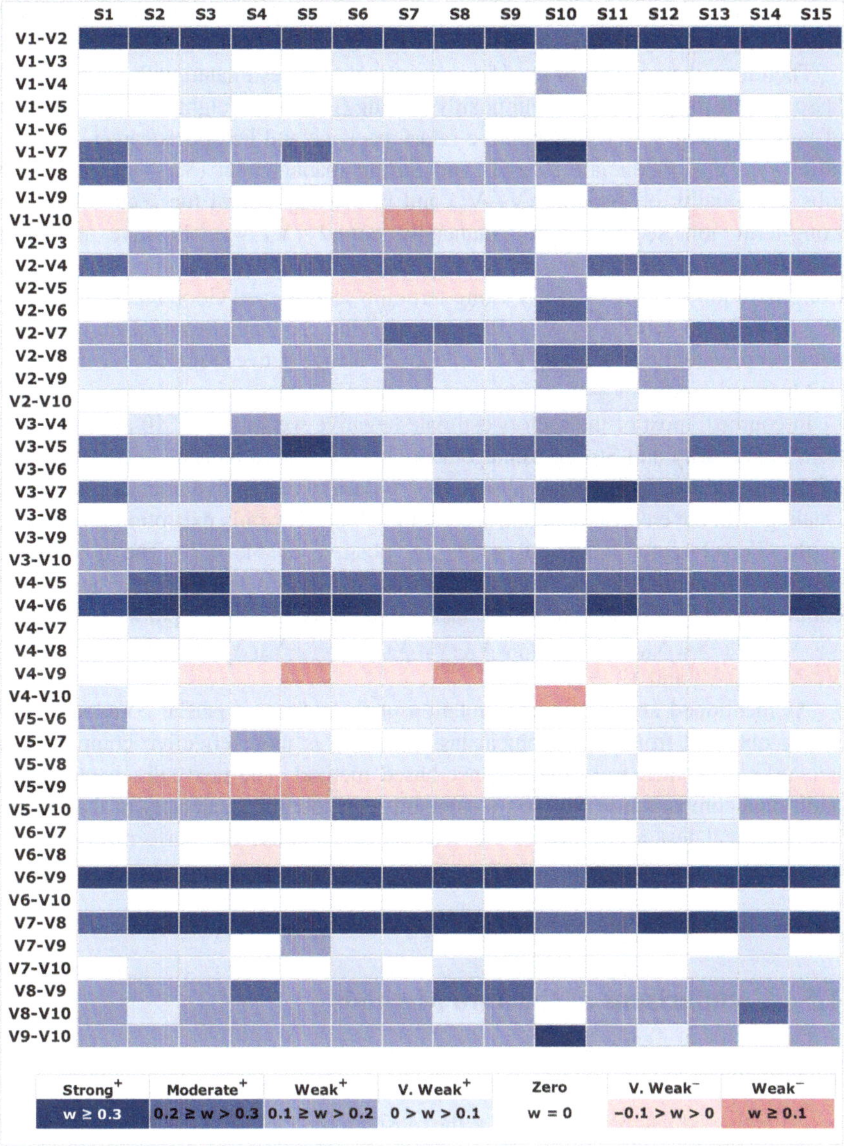

Fig. 4.11 A Correlogram which visualizes the partial correlation coefficients (weights) with EBIC regularization among two variables after conditioning with the remaining observed variables in every sector. Blue color represents positive weights, red color indicates negative weights, while white color denotes the absence of partial correlation between the pair of variables after conditioning on all the other variables in the network

by white color. Moreover, the shade of each rectangle represents the strength of each relationship.

Therefore, the analysis of the 15 graphical networks estimated with the graphical lasso reveals the presence of four positive strong ($w \geq 0.3$) weights between the pair of variables (Fig. 4.11): Competitive salary package and Prospects/career opportunities (V_1–V_2), Financially sounds and Strong management (V_6–V_9), Interesting jobs and Quality of Training (V_7–V_8) and in the majority of the sectors between Long-term job security and Financially sound (V_4–V_6). Positive moderate ($0.2 \leq w < 0.3$) weights are observed between Prospects/career opportunities and Long-term job security (V_2–V_4), Long-term job security and Good balance between private life and work (V_4–V_5), Pleasant working atmosphere and Good balance between private life and work (V_3–V_5), Good balance between private life and work and Interesting jobs (V_3–V_7).

In contrast, most of the sectors estimate negative weights ($w < 0$) among Long-term job security and Strong Management (V_4–V_9), Good balance between private life and work and Strong Management (V_5–V_9) and Competitive salary package and Dealing with the environment and society (V_1–V_{10}). The only exception is observed in the Travel sector where all edges are positive (Table 4.7). The noteworthy common assumption in the majority of these estimated networks is the absence of connectivity ($w = 0$) indicating the lack of causality among 12 pairs of variables, V_2–V_{10}, V_2–V_5, V_6–V_7, V_6–V_{10}, V_6–V_3, V_6–V_8, V_7–V_4, V_7–V_9, V_7–V_4, V_7–V_{10}, V_1–V_4, V_1–V_5 and V_3–V_8.

As mentioned above, the interpretation of the estimated partial correlation networks obtained from the graphical lasso method is more effective comparing to networks calculated from the distance-based method. The next sub-chapter gives additional comparisons of these two methods considering the density of the graphs, the statistical test according to the weights and finally, the performance of the algorithms.

4.5 Comparisons of the Two Methods

A) *Density*

Investigating which method provides the most interpretable graph, Table 4.7 lists the density (δ) of both methods (distance-based and graphical lasso). According to Stach et al. (2012), data-driven networks tend to be denser ($\delta \simeq 0.9$) compared to networks (maps) acquired by humans ($\delta \simeq 0.4$).

Both methods depict the relations among nodes. The distance-based method displayed only complete graphs ($\delta = 1$) with strong positive relationships between nodes together with the low variability. As a result, all edges are displayed as blue and thick (Figs. 4.7 and 4.8). According to research covering the last decade (Haslbeck & Waldorp, 2018; Stach et al., 2012; Scott, 2000), it is difficult to obtain meaningful results from dense graphs. High connectivity leads to more complex and consequently, incomprehensible models increasing the number of spurious connections. As a result, these structures do not represent properly the job-satisfaction problem. On the other hand, Figs. 4.8 and 4.9 illustrate sparser networks with a higher variety of weights between the nodes comparing to distance-based networks. Moreover, the graphical lasso detects the conditional independent edges of partial correlation networks. Since there is no relation between two nodes, it is strongly inferred that causal relationships do not causal relationships between pairs of variables.

This indicates that the density (δ) of each partial correlation network is lower than the distance-based density. Specifically, the density of these networks fluctuates between 0.489 and 0.889 with 0.699 $\pm$ 0.11 (Table 4.7), which is lower than 0.9 of data-driven networks. Moreover, the density of Energy services, Retail, Travel, Automotive and Pharmaceutical job-satisfaction networks are closer to the density ($\delta \simeq 0.4$) of expert-based networks (Stach et al., 2012). Furthermore, graphical lasso networks estimate positive and negative edges, whereas distance-based networks determine only positive edges. Therefore, strong positive connections are visualized as thick and blue (complete opacity), while weak positive connections are indicated as thin and faded blue color. Negative edges are depicted with red color.

B) *Statistical Test*

Next, the investigation of the significant differences of the weighted matrices from distance-based and graphical lasso method are presented. In statistics, to investigate if two samples are significantly different or not, it is necessary first to identify if the population follows the normal distribution and afterwards to select the appropriate statistical test. There are two main approaches to determine if the weights follow the normal distribution relying on statistical tests (numerically) or using visual analysis (graphically).

In this statistical test, objective judgements can be made to assess the normality of the weights. Moreover, due to the small sample size (<50 samples) of both methods, which is equal with the number of possible connections ($n = 45$), the Shapiro-Wilk test is applied to assess the normality of data (Shapiro & Wilk, 1965; Royston, 1982). The null hypothesis (H_0) for this test claims that the two samples represented by the two conditions (methods, weights) are normally distributed. Table 4.8 presents the measurements of Normality Tests for the 15 sectors. To determine if the sample is normally distributed or not, the Sig. (Significance) value plays a decisive role. Generally, the "Sig" or "Sig (2-Tailed)" denotes the p-value. If the Sig. value of the Shapiro-Wilk Test is lower than 0.05, the data significantly vary from the normal distribution. On the contrary, if it is higher than 0.05, the null hypothesis is accepted and there is evidence that the data tested are normally distributed (Shapiro & Wilk, 1965).

In the distance-based method, the data are considered as normal, since the Sig. value of the Shapiro-Wilk Test is higher than 0.05 for all sectors. Furthermore, it is observed that high p-values in Informatics, Non-Profit, Energy services, business and Other indicate that the shapes of the weights are similar to perfect normal distributions ($p_{value} \simeq 1$). Nevertheless, the graphical lasso networks reject the

Table 4.8 Measurements of descriptive statistics, a test of normality and Mann-Whitney (U test) to identify the significance of difference among two observed methods in 15 sectors. DB indicates distance-based and GL signifies graphical lasso method

		Descriptives							Shapiro-Wilk		Test statistics		
	Methods	Min	Max	Std. Dev	Range	Variance	Skewness	Kurtosis	Statistic	Sig.	Mann-Whitney U	Z	Asymp. Sig. (2-tailed)
Automotive	DB	0.742	0.849	0.026	0.107	0.001	0.358	−0.574	.966	.215	0	−8.19	.000
	GL	−0.097	0.340	0.109	0.437	0.012	0.600	−0.546	.899	.001			
Constructor	DB	0.749	0.844	0.023	0.095	0.001	0.145	−0.401	.985	.821	0	−8.17	.000
	GL	−0.103	0.388	0.105	0.490	0.011	0.981	0.760	.907	.002			
Finance	DB	0.725	0.863	0.029	0.138	0.001	−0.133	0.077	.986	.867	0	−8.17	.000
	GL	−0.123	0.378	0.123	0.501	0.015	0.687	0.162	.939	.020			
Food	DB	0.745	0.848	0.023	0.103	0.001	−0.134	−0.001	.987	.878	0	−8.17	.000
	GL	−0.113	0.349	0.118	0.462	0.014	0.507	−0.578	.941	.023			
HR	DB	0.738	0.853	0.027	0.115	0.001	0.056	−0.291	.989	.942	0	−8.19	.000
	GL	−0.154	0.414	0.129	0.568	0.017	0.738	0.173	.919	.004			
Industry	DB	0.739	0.859	0.026	0.120	0.001	0.236	−0.090	.990	.965	0	−8.18	.000
	GL	−0.073	0.383	0.116	0.456	0.013	0.852	0.077	.916	.003			
Informatics	DB	0.724	0.858	0.029	0.134	0.001	−0.042	−0.159	.991	.985	0	−8.17	.000
	GL	−0.110	0.406	0.121	0.515	0.015	0.813	0.403	.937	.017			
Non-profit	DB	0.726	0.856	0.029	0.130	0.001	−0.065	−0.154	.995	.999	0	−8.17	.000
	GL	−0.139	0.428	0.130	0.567	0.017	0.801	0.304	.939	.020			
Pharmaceutical	DB	0.749	0.855	0.024	0.106	0.001	0.248	−0.453	.982	.694	0	−8.19	.000
	GL	−0.038	0.411	0.117	0.449	0.014	1.141	0.646	.848	.000			
Energy services	DB	0.773	0.857	0.020	0.084	0.000	−0.036	−0.240	.988	.925	0	−8.24	.000
	GL	−0.174	0.332	0.125	0.506	0.016	0.337	−1.011	.839	.000			
Retail	DB	0.727	0.848	0.029	0.121	0.001	−0.187	−0.251	.981	.670	0	−8.22	.000
	GL	−0.090	0.411	0.124	0.501	0.015	1.018	0.195	.840	.000			

Business	DB	0.746	0.860	0.026	0.114	0.001	0.188	−0.385	.989	.949	0	−8.18	.000
	GL	−0.073	0.392	0.113	0.465	0.013	0.968	0.377	.904	.000			
Transportation	DB	0.758	0.864	0.025	0.106	0.001	0.348	−0.401	.970	.303	0	−8.18	.000
	GL	−0.060	0.500	0.118	0.559	0.014	1.231	1.641	.885	.000			
Travel	DB	0.734	0.857	0.027	0.123	0.001	−0.134	0.145	.977	.516	0	−8.20	.000
	GL	0.000	0.411	0.116	0.411	0.013	1.186	0.566	.815	.000			
Other	DB	0.741	0.855	0.027	0.114	0.001	0.112	−0.398	.986	.849	0	−8.17	.000
	GL	−0.059	0.453	0.118	0.512	0.014	1.196	0.871	.880	.000			

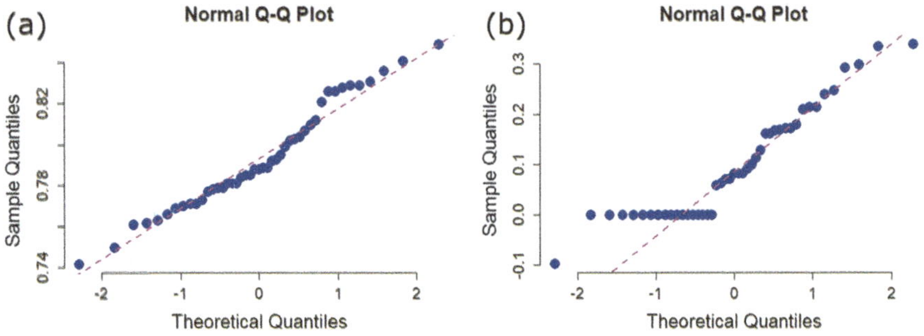

Fig. 4.12 The Normal Q-Q Plots examine the normal distribution of the weight coefficients of the (**a**) distance-based and the (**b**) graphical-lasso methods in the Automotive sector (S_1), respectively. The dashes straight line denotes the normal distribution in the Q-Q plots

null hypothesis since the weights are not normally distributed because the p-values are lower than 0.05.

Remaining in statistical perspective, one graphical approach is the visualization of Normal Q-Q Plots to analyze and compare two probability distributions by plotting their quantiles against each other. Figure 4.12 illustrates the samples of the weight matrices of the distance-based and glasso algorithm in the Automotive sector. In the distance-based method (Fig. 4.12a), the two compared distributions are exactly equal since the points on the Q-Q plot are perfectly lie on a dashed straight line ($y = x$). On the contrary, the data points in the graphical lasso are depicted far from the line (in an obvious non-linear way) indicating that the weights are not normally distributed (Fig. 4.12b).

Thus, Fig. 4.13 demonstrates the second graphical approach of the distributions of the weights of both methods for the 15 sectors. It is noted that in the distance-based method the weights are normally distributed (red color), but in the graphical lasso, the weights do not follow the normal distribution curve (blue color). In particular, the distribution of the graphical lasso is right-skewed (which is also known as positive skew in which the tail on the right side of the probability density function is longer than the left side). In this case, the distribution follows the exponential distribution.

To summarize, the test of normality shown that the samples of the graphical lasso do not follow the "bell" shape. For this reason, the Mann-Whitney U test (Mann & Whitney, 1947) is examined as the appropriate statistical test to identify if the observed methods (independent groups) are significantly different (H_1) or not (H_0). Thus, the Mann-Whitney U test compares the number of times a score from one sample is ranked higher than a score from another sample. If the scores of the samples are equal, a tie is determined. In general, the Mann-Whitney test is symbolized by U. When the sample size is large enough, the determination of U becomes long. Additionally, if the observed value of U is lower than the critical value ($U_{crit.\ v}$) then the test is significant (at the α level) and the null hypothesis is rejected. Moreover, measurements regarding statistic test, U statistic, z-value and the

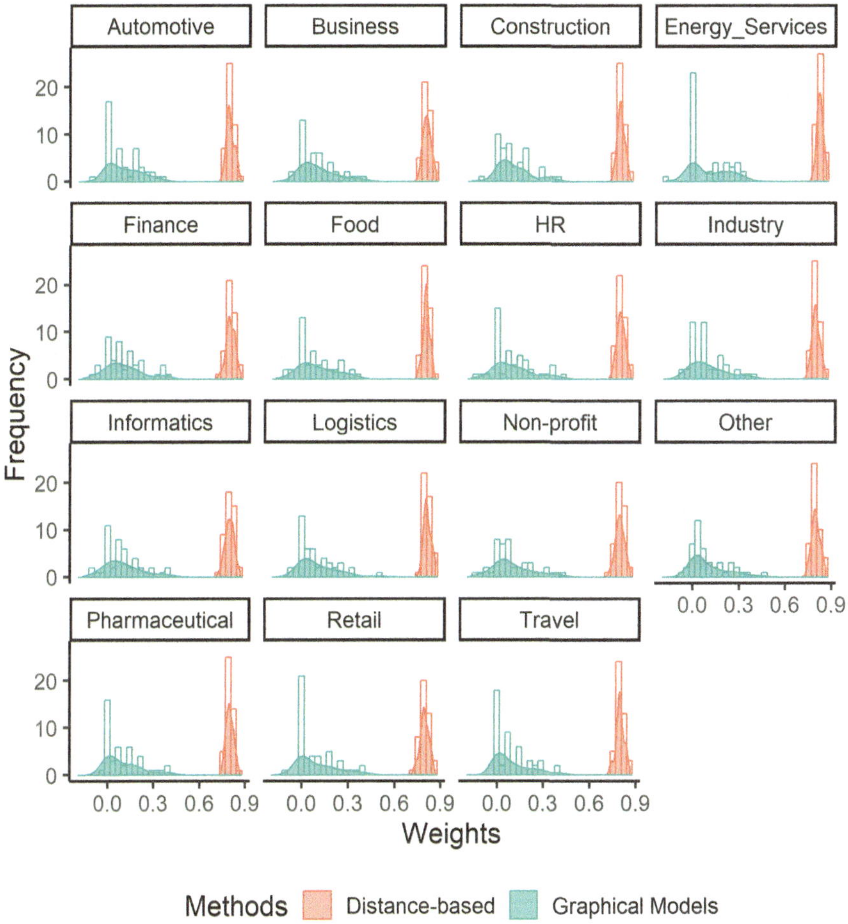

Fig. 4.13 Distributions of 15 sectors indicate that the weights derived from the distance-based are normally distributed; while, in Graphical Models, all distributions are left-skewed

p-value (or the 2-tailed asymptotic significance) are additionally provided. If the absolute value of the obtained z-score is less than the critical value of 1.96, then H_0 is retained. (*H_0: The two populations represented by the two conditions (methods, weights) have the same distribution of scores*); while the absolute value of the obtained z-score is greater than 1.96, then H_0 is rejected. Additionally, two groups are considered significantly different when the p-value is lower than 0.05 ($p < 0.05$).

Regarding the observed distributions of the weights of distance-based and graphical lasso method have different shapes (Fig. 4.13), the Mann-Whitney U test is using to compare the mean ranks (Table 4.8) of the dependent variable (weights) and both the independent variables (methods). As it is expected, the mean ranks of the graphical lasso are lower than the distance-based, since it estimates lower weights. Thus, the asymptotic significance (2-tailed) or p-value is equal to zero and the

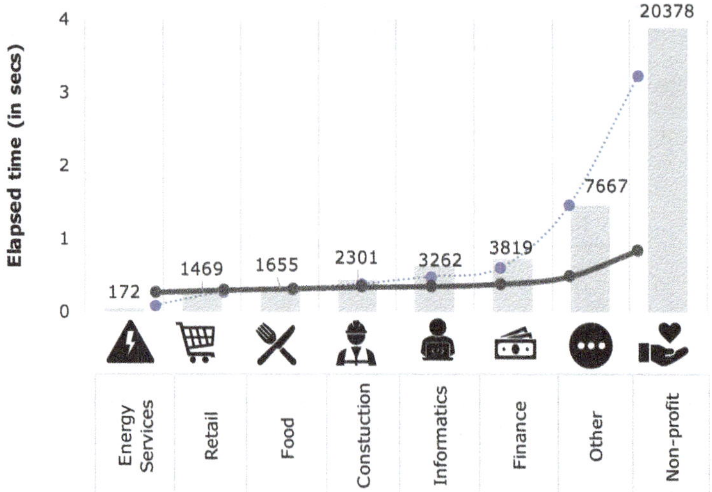

	Energy Services	Retail	Food	Constuction	Informatics	Finance	Other	Non-profit
Observations	172	1469	1655	2301	3262	3819	7667	20378
Distance-based	0.09	0.28	0.32	0.39	0.49	0.61	1.47	3.24
Graphical Lasso	0.27	0.3	0.32	0.34	0.35	0.38	0.48	0.83

Fig. 4.14 The estimation of elapsed time in seconds of the distance-based and the graphical lasso method

absolute value of the obtained z-score is greater than the critical value 1.96, for all sectors. This concludes that the distributions of the two methods are significantly different ($U = 0.000$, $N = 45$, $p - value < 0.05$ two-tailed and $z > 1.96$). Consequently, the H_1 hypothesis is retained, while the H_0 is rejected. For the informational purpose, if two distributions considered as similar then the medians of the dependent variable are compared.

C) *Performance of two algorithms*

The objectives of this chapter are not only to compare the density and the variations of weights between the structures of each sector but also to estimate the performance of both algorithms determining the elapsed time. In other words, it is assessed the amount of time that CPU (Central Processing Unit) requires to process the R session. Therefore, we have implemented in the R programming language the distance-based method to estimate the elapsed time and identify which of the two methods is less time-consuming. The function *proc.time()* was used to figure out the CPU time (in seconds). It is observed that the graphical lasso performed efficiently in larger datasets and the variation of elapsed time between two methods is getting higher as the population of the dataset increases (Fig. 4.14).

Specifically, the black line represents the execution time of the graphical lasso algorithm. This line remains constant as the number of observations is lower than 4000, while a slight increase is noticed when the datasets are very large (Other and Non-Profit sectors). Particularly, in Finance, the execution time of the graphical lasso is 1.6 times smaller than the distance-based method; in Other sector, the execution time is three times smaller and in Non-profit, the difference between two methods is quadruple. Nevertheless, the performances of the graphical lasso algorithm are least advantageous when the datasets are very small. Particularly, the elapsed time of graphical lasso is three times greater than the distance-based method. Moreover, the time in Retail, Food and Constructor is not significantly different. Notwithstanding these drawbacks, the graphical lasso line is characterized as linear with a steady increase, while the distance-based (dotted blue) line follows the exponential growth indicating that in larger datasets the time is increasing rapidly.

To sum up, the complexity of the glasso algorithm (Friedman et al., 2008; Mazumder & Hastie, 2012) using a row-by-row block coordinate method is roughly $O(p^3)$ for reasonably sparse-problems with p nodes. For high-dimensional problems, the cost can be as large as $O(p^4)$. However, Haslbeck and Waldorp (2020) optimized the computational complexity of the algorithm into $O(p \log(2 \cdot p))$ using the nodewise regression approach. On the other hand, we calculate the time complexity of the distance-based method according to our programming code in R and it is equal to $O(m \cdot p^2)$, m denotes the number of observations. Even if, we can optimize the distance-based method, we confirmed that the efficiency of the distance-based method in large datasets (over 1000) is poor since the created models are fully connected and therefore, are hard to interpret.

4.6 Conclusions

The major goal of the current chapter was to construct a conceptual model from cross-sectional data that describes how the observed features associated with a job-satisfaction problem would interact within a larger model. The sample of this study was over than 50,000 observations which 10 variables associated with the job satisfaction were evaluated in a five-point Likert scale. Each observation (row) corresponds to an employee who was occupied in 1 of the 15 working sectors in Belgium.

First, we briefly described the definition of a network and its classifications. Then, we investigated if the variables were clustering in different groups to employ a dimensionality reduction technique to decrease the dimension of our feature set. For this reason, nine methods of phycho R package (Makowski, 2018) were performed to identify the possible clusters of our datasets. The majority of the procedures (66.7% or six out of nine) indicated that the variables of the 15 samples were clustering all together in one cluster. Therefore, the qgraph R package (Epskamp et al., 2012) was utilized a well-known technique from machine learning, the lasso regularization with the EBIC model selection for network modeling. Particularly, the EBIC adds an extra penalty, the hyperparameter γ to control the model complexity. Then, we compared the distance-based method with the glasso algorithm. Results showed that graphical lasso estimated sparser graphs easier to interpret with efficient performance (mainly in dataset over of 2000 observations) comparing to the distance-based method (Schneider et al., 1998) that derives complete graphs with positive edges, hard to interpret and deficient performance (higher time performance) to large datasets. Specifically, the density of 15 graphical models fluctuates between 0.49 and 0.89, while the connections were positive and negative. Even if, we can optimize the distance-based method, it is impossible to increase the efficiency of the distance-based method since the created models are fully connected and therefore, are hard to interpret (Dikopoulou et al., 2017).

References

Ahmadi, S., Papageorgiou, E., Yeh, C.-H., & Martin, R. (2015). Managing readiness-relevant activities for the organizational dimension of ERP implementation. *Computers in Industry, 68*, 89–104.

Akerkar, R., & Lingras, P. (2007). *Building an intelligent web: Theory and practice* (1st ed., pp. 177–189). Jones and Bartlett Publishers.

Alsberg, B. K. (2000). Wavelets in parsimonious functional data analysis models. *Wavelets in Chemistry*, 351–410. https://doi.org/10.1016/s0922-3487(00)80041-6

Arvan, M., Omidvar, A., & Ghodsi, R. (2016). Intellectual capital evaluation using fuzzy cognitive maps: A scenario-based development planning. *Expert Systems with Applications, 55*, 21–36.

Bakhtavar, E., & Shirvand, Y. (2018). Designing a fuzzy cognitive map to evaluate drilling and blasting problems of the tunneling projects in Iran. *Engineering with Computers*. https://doi.org/10.1007/s00366-018-0581-y

Banerjee, S., Gelfand, A. E., Finley, A. O., & Sang, H. (2008). Gaussian predictive process models for large spatial data sets. *Journal of the Royal Statistical Society: Series B (Statistical Methodology), 70*(4), 825–848. https://doi.org/10.1111/j.1467-9868.2008.00663.x

Bhoj, D. S., & Ahsanullah, M. (1994). Estimation of a conditional mean in a linear regression model after a preliminary test on regression coefficient. *Biometrical Journal, 36*(2), 153–163. https://doi.org/10.1002/bimj.4710360206

Boyd, S., & Vandenberghe, L. (2004). *Convex optimization*. Cambridge University Press.

Braeken, J., & van Assen, M. A. L. M. (2017). An empirical Kaiser criterion. *Psychological Methods, 22*(3), 450–466.

Chen, J., & Chen, Z. (2008). Extended bayesian information criteria for model selection with large model spaces. *Biometrika, 95*(3), 759–771.

Clifford, P. (1990). Markov random fields in statistics. In G. R. Grimmett & D. J. A. Welsh (Eds.), *Disorder in physical systems* (pp. 19–32). A Volume in Honour of John M. Hammersley, Clarendon Press.

Costantini, G., Epskamp, S., Borsboom, D., Perugini, M., Mõttus, R., Waldorp, L. J., & Cramer, A. O. J. (2015). State of the aRt personality research: A tutorial on network analysis of personality data in R. *Journal of Research in Personality, 54*, 13–29.

Cox, D. R., & Wermuth, N. (1993). Linear dependencies represented by chain graphs. *Statistical Science, 8*(3), 204–218.

Cramer, A. O. J., van der Sluis, S., Noordhof, A., Wichers, M., Geschwind, N., & Aggen, S. H. (2012). Dimensions of normal personality as networks in search of equilibrium: You can't like parties if you don't like people. *European Journal of Personality, 26*(4), 414–431. https://doi.org/10.1002/per.1866

Dahlin, M., Joneborg, N., & Runeson, B. (2005). Stress and depression among medical students: A cross sectional study. *Medical Education, 39*(6), 594–604.

De Ron, J., Fried, E. I., & Epskamp, S. (2019). Psychological networks in clinical populations: Investigating the consequences of Berkson's bias. *Psychological Medicine*.

Dikopoulou, Z., Papageorgiou, E., Mago, V., & Vanhoof, K. (2017). A new approach using mixed graphical model for automatic design of fuzzy cognitive maps from ordinal data. *IEEE International Conference on Fuzzy Systems (FUZZ-IEEE)*. https://doi.org/10.1109/fuzz-ieee.2017.8015607

Dikopoulou, Z., Papageorgiou, E., & Vanhoof, K. (2020). From undirected structures to directed graphical lasso fuzzy cognitive maps using ranking-based approaches. IEEE International Conference on Fuzzy Systems (FUZZ-IEEE).

Drton, M., & Perlman, M. D. (2004). Model selection for Gaussian concentration graphs. *Biometrika, 91*(3), 591–602.

Epskamp, S., & Fried, E. I. (2017). A tutorial on regularized partial correlation networks. *Psychological Methods*. In press. https://doi.org/10.1037/met0000167

Epskamp, S. (2016). *Regularized Gaussian psychological networks: Brief report on the performance of extended BIC model selection*. arXiv preprint, arXiv:1606.05771.

Epskamp, S., Borsboom, D., & Fried, E. I. (2017a). Estimating psychological networks and their accuracy: A tutorial paper. *Behaviour Research Methods, 2017*, 1–18.

Epskamp, S., Cramer, A., Waldorp, L., Schmittmann, V. D., & Borsboom, D. (2012). qgraph: Network visualizations of relationships in psychometric data. *Journal of Statistical Software, 48* (4), 1–18. http://www.jstatsoft.org/v48/i04/

Epskamp, S., Kruis, J., & Marsman, M. (2017b). Estimating psychopathological networks: Be careful what you wish for. *PLoS One, 12*(6), e0179891. https://doi.org/10.1371/journal.pone.0179891

Epskamp, S., Waldorp, L. J., Mõttus, R., & Borsboom, D. (2018). The Gaussian graphical model in cross-sectional and time-series data. *Multivariate Behavioral Research, 53*(4), 453–480. https://doi.org/10.1080/00273171.2018.1454823

Foygel, R., & Drton, M. (2010). Extended Bayesian information criteria for Gaussian graphical models. *Advances in Neural Information Processing Systems, 23*, 2020–2028.

Fried, E. I., Epskamp, S., Nesse, R. M., Tuerlinckx, F., & Borsboom, D. (2016). What are 'good' depression symptoms? Comparing the centrality of DSM and non-DSM symptoms of depression in a network analysis. *Journal of Affective Disorders, 189*, 314–320.

Friedman, J. H., Hastie, T., & Tibshirani, R. (2008). Sparse inverse covariance estimation with the graphical lasso. *Biostatistics, 9*(3), 432–441.

Friedman, J. H., Hastie, T., & Tibshirani, R. (2014). glasso: Graphical lasso estimation of gaussian graphical models [Computer software manual]. https://CRAN.R-project.org/package=glasso

Fruchterman, T. M., & Reingold, E. M. (1991). Graph drawing by force-directed placement. *Software: Practice and Experience, 21*, 1129–1164.

Grover, S., Dua, D., Chakrabarti, S., & Avasthi, A. (2018). Factor analysis of symptom dimensions (psychotic, affective and obsessive compulsive symptoms) in schizophrenia. *Asian Journal of Psychiatry, 38*, 72–77. https://doi.org/10.1016/j.ajp.2017.10.028

Hammersley, J. M., & Clifford, P. (1971). *Markov field on finite graphs and lattices*. Unpublished.

Hartigan, J. A., & Wong, M. A. (1979). Algorithm AS 136: A K-means clustering algorithm. *Applied Statistics, 28*(1), 100–108.

Haslbeck, J. M. B., & Waldorp, L. J. (2018). How well do network models predict future observations? On the importance of predictability in network models. *Behavior Research Methods, 50*(2), 853–861. https://doi.org/10.3758/s13428-017-0910-x

Haslbeck, J. M. B., & Waldorp, L. J. (2020). mgm: Estimating time-varying mixed graphical models in high-dimensional data. *Journal of Statistical Software, 93*(8). https://doi.org/10.18637/jss.v093.i08

Hastie, T., Tibshirani, R., & Friedman, J. H. (2009). *The elements of statistical learning: Data mining, inference, and prediction* (2nd ed.). Springer.

Hastie, T., Tibshirani, R., & Wainwright, M. (2015). *Statistical learning with sparsity: The lasso and generalizations*.

Isvoranu, A. M., Guloksuz, S., Epskamp, S., van Os, J., Borsboom, D., & GROUP. (2019). Towards incorporating genetic risk scores into symptom networks of psychosis. *Psychological Medicine.*

Jensen, S. T., Johansen, S., & Lauritzen, S. L. (1991). Globally convergent algorithms for maximizing likelihood function. *Biometrika, 78*(4), 867. https://doi.org/10.2307/2336939

Johnson, S. C. (1967). Hierarchical clustering schemes. *Psychometrika, 32*(3), 241–254. https://doi.org/10.1007/bf02289588

Kaiser, H. F. (1960). The application of electronic computers to factor analysis. *Educational and Psychological Measurement, 20*(1), 141–151.

Kassambara, A., & Mundt, F. (2017). *Extract and visualize the results of multivariate data analyses*. CRAN. https://cran.rproject.org/web/packages/factoextra/factoextra.pdf

Knowland, V. C. P., Purser, H., & Thomas, M. S. C. (2015). Cross-sectional methodologies in developmental psychology. *International Encyclopedia of the Social & Behavioral Sciences*, 354–360. https://doi.org/10.1016/B978-0-08-097086-8.23235-2

Koller, D., & Friedman, N. (2009). *Probabilistic graphical models: Principles and techniques*. MIT Press.

Kossakowski, J. J., Epskamp, S., Kieffer, J. M., van Borkulo, C. D., Rhemtulla, M., & Borsboom, D. (2015). The application of a network approach to health-related quality of life (HRQoL): Introducing a new method for assessing HRQoL in healthy adults and cancer patient. *Quality of Life Research, 25*, 781–792.

Kotu, V., & Deshpande, B. (2019). Clustering. *Data Science*, 221–261. https://doi.org/10.1016/b978-0-12-814761-0.00007-1

Lauritzen, S. L. (1996). *Graphical models*. Clarendon Press.

Levin, K. A. (2006). Study design III: Cross-sectional studies. *Evidence-Based Dentistry, 7*, 24–25. https://doi.org/10.1038/sj.ebd.6400375

Lloyd, S. (1982). Least squares quantization in PCM. *IEEE Transactions on Information Theory, 28*, 129–137.

Loh, L., & Venkatraman, N. (1992). Determinants of information technology outsourcing: A cross-sectional analysis. *Journal of Management Information Systems, 9*(1), 7–24. https://doi.org/10.1080/07421222.1992.11517945

Maechler, M., Rousseeuw, P., Struyf, A., Hubert, M., Hornik, K., Studer, M., Roudier, P., Gonzalez, J., Kozlowski, K., & Schubert, E. (2019). *Finding groups in data*. CRAN. https://cran.r-project.org/web/packages/cluster/cluster.pdf

Mair, P. (2018). *Factor analysis. Modern psychometrics with R* (pp. 17–61). Springer.

Makowski, D. (2018). The psycho package: An efficient and publishing-oriented workflow for psychological science. *Journal of Open Source Software, 3*(22), 470. https://doi.org/10.21105/joss.00470

Mann, B. H., & Whitney, R. D. (1947). On a test of whether one of two random variables is stochastically larger than the other. *Annals of Mathematical Statistics, 18*(1), 50–60.

Mazumder, R., & Hastie, T. (2012). The graphical lasso: New insights and alternatives. *Electronic Journal of Statistics, 6*, 2125–2149.

McArdle, J. J., & Kadlec, K. M. (2013). Structural equation models. In T. D. Little (Ed.), *The Oxford handbook of quantitative methods* (Statistical analysis) (Vol. 2, pp. 295–337). Oxford University Press.

McNally, R. J. (2016). Can network analysis transform psychopathology? *Behaviour Research and Therapy, 86*, 95–104. https://doi.org/10.1016/j.brat.2016.06.006

Meinshausen, N., & Bühlmann, P. (2006). High-dimensional graphs and variable selection with the lasso. *The Annals of Statistics*, 1436–1462.

Mike, L., Guimond, S., Kelly, S., Thermenos, H., Mesholam-Gately, R., Eack, S., & Keshavan, M. (2019). Social cognition in early course of schizophrenia: Exploratory factor analysis. *Psychiatry Research, 272*, 737–743. https://doi.org/10.1016/j.psychres.2018.12.152

Murphy, K. P. (2012). *Machine learning: A probabilistic perspective*. MIT press.

Muthén, B. (1984). A general structural equation model with dichotomous, ordered categorical, and continuous latent variable indicators. *Psychometrika, 49*, 115–132. https://doi.org/10.1007/BF02294210

Newman, M. E. J. (2010). *Networks: an introduction* (pp. 3, 5, 6, 34, 77, 195, 205, 208). Oxford University Press.

Norris, M., & Lecavalier, L. (2009). Evaluating the use of exploratory factor analysis in developmental disability psychological research. *Journal of Autism and Developmental Disorders., 40* (1), 8–20. https://doi.org/10.1007/s10803-009-0816-2

Olsson, U. (1979). Maximum likelihood estimation of the polychoric correlation coefficient. *Psychometrika, 44*, 443–460. https://doi.org/10.1007/BF02296207

Pearson, K. (1900). Mathematical contributions to the theory of evolution. VII. On the correlation of characters not quantitatively measurable. *Philosophical Transactions of the Royal Society of London. Series A: Mathematical, Physical and Engineering Sciences, 195*(1), 47.

Poghosyan, L., Aiken, L. H., & Sloane, D. M. (2009). Factor structure of the Maslach burnout inventory: An analysis of data from large scale cross-sectional surveys of nurses from eight countries. *International Journal of Nursing Studies, 46*, 894–902.

Raykov, T., & Marcoulides, G. A. (1999). On desirability of parsimony in structural equation model selection. *Structural Equation Modeling: A Multidisciplinary Journal, 6*(3), 292–300. https://doi.org/10.1080/10705519909540135

Rodriguez-Repiso, L., Setchi, R., & Salmeron, J. L. (2007). Modeling IT projects success with fuzzy cognitive maps. *Expert Systems with Applications, 32*(2), 543–559.

Royston, J. P. (1982). An extension of Shapiro and Wilk's W test for normality to large samples. *Applied Statistics, 31*, 115–124.

Schneider, M., Shnaider, E., Kandel, A., & Chew, G. (1998). Automatic construction of FCMs. *Fuzzy Sets and Systems, 93*(2), 161–172.

Scott, J. (2000). *Social network analysis: A handbook* (2nd ed.). Sage.

Shapiro, S. S., & Wilk, M. B. (1965). An analysis of variance test for normality (complete samples). *Biometrika, 52*(3–4), 591–611.

Stach, W., Pedrycz, W., & Kurgan, L. A. (2012). Learning of fuzzy cognitive maps using density estimate. *IEEE Transactions on Systems, Man, and Cybernetics. Part B, 42*(3), 900–912.

Stylios, C. D., Bourgani, E., & Georgopoulos, V. C. (2020). Impact and applications of fuzzy cognitive map, methodologies. In O. Kosheleva, S. P. Shary, G. Xiang, & R. Zapatrin (Eds.), *Beyond traditional probabilistic data processing techniques: Interval, fuzzy etc. Methods and their applications. Studies in computational intelligence* (pp. 229–246).

Thompson, B. (2007). Factor analysis. *The Blackwell Encyclopedia of Sociology*. https://doi.org/10.1002/9781405165518.wbeosf003

Thurstone, L. L. (1931). Multiple factor analysis. *Psychological Review, 38*(5), 406–427.

Tibshirani, R. (1996). Regression shrinkage and selection via the lasso. *Journal of the Royal Statistical Society. Series B (Methodological), 58*, 267–288.

van Borkulo, C. D., Boschloo, L., Borsboom, D., Penninx, B. W. J. H., Waldorp, L. J., & Schoevers, R. A. (2015). Association of symptom network structure with the course of depression. *JAMA Psychiatry, 72*(12), 1219–1226.

Vazigiannis, M. (2003). Data modeling: Object-oriented data model. *Encyclopedia of Information Systems*, 505–320. https://doi.org/10.1016/B0-12-227240-4/00035-6

Wasserman, L. (2004). All of statistics. *Springer Texts in Statistics*. https://doi.org/10.1007/978-0-387-21736-9

Yuan, M., & Lin, Y. (2007). Model selection and estimation in the Gaussian graphical model. *Biometrika, 94*, 19–35.

Zarrin, M., & Azadeh, A. (2018). Mapping the influences of resilience engineering on health, safety, and environment and ergonomics management system by using Z-number cognitive map. *Human Factors and Ergonomics in Manufacturing & Service Industries*. https://doi.org/10.1002/hfm.20766

Zhao, P., & Yu, B. (2006). On model selection consistency of lasso. *The Journal of Machine Learning Research, 7*, 2541–2563.

Chapter 5
Network Analysis, Accuracy and Stability of the Job-Satisfaction Structures

The network structure is defined as highly interconnected systems, which describe various contexts such as technological and transportation infrastructure, social phenomena, biological systems, ecological systems, etc. Furthermore, any structured technique that is used to analyze mathematically a network of interconnected components and characterize the type of the network is referred to as "*network analysis*". Consequently, different network methods provide alternative approaches to analyze data and suggest new alternatives for modelling and simulating job attractiveness processes. In this chapter, an overview of network analysis approaches is presented to specify how to determine and interpret meaningful measures of local and global network indices. The observed networks were constructed applying the lasso regularization technique as described in Chap. 4. Finally, we investigate how bootstrapping routines are applied to assess stability and the stability of the network inferences. In particular, we use a non-parametric bootstrap method (called as "*resampled with replacement*") to access the accuracy of the networks' parameters and bootstrap based on subsets of the data (known also as "*case-dropping subset bootstrap*") for the network stability. Results show that larger datasets provide more accurate and stable models since the mean bootstrapped CIs of edge-weights are not significantly differed from the actual weights and the Correlation Stability coefficient (CS-coefficient) that measures the 95% of the confidence intervals of the correlations is above 0.5.

5.1 Introduction

Numerous systems in real-world can be defined by models of complex networks such as the Internet consisting of routers, the brain consisting of neurons, social networks consisting of people or even more complex systems that many parts are distributed without centralized control (Newman, 2010; Strogatz, 2001). However, as Chap. 4 describes, it is critical for the estimated models to avoid overfitting.

Z. Dikopoulou, *Modeling and Simulating Complex Business Perceptions*, Fuzzy Management Methods, https://doi.org/10.1007/978-3-030-81496-0_5

Specifically, using the glasso method with regularization, the most convenient process to minimize the overfitting is to shrink small coefficients exactly to zero. Therefore, glasso method with the EBIC regularization eliminates spurious edges and consequently, the interpretability is increased since sparser models are obtained. Another alternative to minimize overfitting is to use larger dataset because the estimated model is highly accurate (Murphy, 2012). Nevertheless, in smaller datasets, the network parameters may not be accurate. For this reason, it is necessary to assess the accuracy of these networks (a) using bootstrapped confidence intervals of estimated edge-weights and (b) investigating the stability of centrality indices after observing subsets of the data.

The aim of this contribution is the description and the application of the necessary theoretical and methodological tools to analyze the networks associated with the job satisfaction problem. Network analysis measurements are applied to identify the local and the global features of the networks. Specifically, this chapter introduces many inference methods adapted from the graph theory to determine the most important (central) nodes in the network. Afterwards, additional basic properties of complex networks like the average path length, the degree distribution and the clustering coefficient to distinguish the structure of the network (Newman, 2001, 2010). All measurements concerning the inference and the structure of job satisfaction networks are performed using libraries qgraph (Epskamp et al., 2012) and igraph (Csardi & Nepusz, 2006) implemented in R programming language. Finally, in order to investigate how accurate are the weights of each job satisfaction network, the bootnet R package (Epskamp et al., 2017) is applied to determine the accuracy of business networks by estimating the confidence intervals (CIs) on the edge-weights and the stability of the centrality indices using the case-dropping subset bootstrap.

5.2 Centrality Measurement

On the estimated network, not all nodes are equally important; therefore, different measurements are necessary to infer the most important nodes in the network (Kolaczyk, 2009a). The importance of each node in the network can be assessed by evaluating the node's centrality. The conception of centrality is manifold in which each centrality measurement describes a specific type of centrality. Every type is based on the patterns of the connections that constitute a network. Principally, a node is characterized as central if it strongly and directly influences many other nodes, if it has direct and indirect shorter paths that connects itself to other nodes and if it is regularly detected on the shortest path among two nodes. A node that is not central is called "*peripheral*".

Consequently, several indices of centrality have been proposed (Freeman, 1978; Sabidussi, 1966) to identify which nodes are more central. This subchapter presents the most prominent centrality measures such as degree, strength, betweenness and closeness.

5.2.1 Network Inference

After the estimation of the data-driven network (Chap. 4), the visual inspection can be a first effortless step to interpret the relative importance of nodes (Gramer et al., 2012). For small and sparse networks, an effective algorithm to clarify connectivity of the network is the Fruchterman-Reingold's algorithm (Fruchterman & Reingold, 1991). This algorithm locates the nodes of the network according to the link weights. Specifically, strongly connected nodes are placed closer; while weakly related nodes are set away from each other. However, for larger networks, an approach from graph theory named network analysis is an appropriate procedure to infer the local and global properties of the networks related to the business problem.

The graphical representation of the derived network is considered as an initial step of network analysis to show the structure of the data (Epskamp et al., 2012). Therefore, inference methods adapted from graph theory can be applied to analyse the network structure, the global and the local structure. The global structure answers questions about the network as a whole entity (Humphries & Gumey, 2008; Newman, 2001; Watts & Strogatz, 1998; Luce & Perry, 1949). How large is it? How dense is it? Does it obtain any strong clusters? On the other hand, local structure addresses different features for each node which is useful to comprehend how nodes differ (Newman, 2001, 2010; Borgatti & Everett, 2006; Freeman, 1978). Which of the nodes is the most central? Which node has the strongest connections? What is the shortest path between two nodes?

Applying the lasso regularization technique to the data, the estimated networks are always weighted undirected networks. Consequently, many typical inference methods related to the global structure indices, such as density, global clustering and small-worldness are not that meaningful to describe properly the job satisfaction networks (Boccaletti et al., 2006; Kolaczyk, 2009b; Newman, 2010). The reason is that these measurements take into consideration the presence or the absence of connections among nodes and not the significance of the weights that every connection carries on. Therefore, this chapter describes and analyses the measures of the relevant job satisfaction networks, revolving mostly around the centrality and the cluster coefficients indices.

5.2.1.1 Degree and Strength

Degree centrality is deemed to be the simplest node centrality index. It is estimated by using the local structure and it is defined as the number of edges that are connected to the observed node (Freeman, 1978). This measure has been formalized as follows: $k_i = \sum_j a_{ij}$, where i denotes the node of interest, j includes the remaining nodes of the network and a_{ij} represents the adjacency matrix. Thus, for every node i, the ratio between its degree and the number of nodes in the network minus one is defined as degree centrality, $C_{\deg}(i) = k_i/(|N| - 1)$, where N represents the total number of nodes. Consequently, a node would have a degree centrality of one if it is connected to every single node in the network and zero if it is isolated indicating that it is not connected to other nodes (Freeman, 1978). Moreover, the average degree of the network, $\langle k \rangle$ is considered as the average degrees k_i over the total number of nodes (N).

Another important measurement that captures a small amount of information about the structure of a network is the degree distribution $P_{\deg(k)}$. The degree distribution $P_{\deg(k)}$ is the probability that a randomly selected node has k connections (or neighbours) and it is defined by the fraction of nodes in the graph with degree k. For instance, in Fig. 4.1a, the highest degree of node C is $k_C = 4$, since it has four neighbours, node A, B, D and E. Therefore, the degree centrality of node C is 1, $C_{\deg(C)} = 4/(5 - 1)$ and the degree distribution of the network is $P_{\deg(1)} = 1/4$, $P_{\deg(2)} = 2/4$, $P_{\deg(3)} = 1/4$, $P_{\deg(4)} = 1/4$ and all other $P_{\deg(k)} = 0$.

As a generalization of the degree index could be another index that describes the nodes in weighted networks. This generalization is labelled as node strength (Barrat et al., 2004; Newman, 2004). Strength is defined by the absolute summation of weights of node i and it is formalized as $s_i = \sum_j |w_{ij}|$, where w_{ij} represents the link of the weighted adjacency matrix. Therefore, the strength of node C in Fig. 4.1b is $s_C = 0.8$. As it is observed, even if more neighbours are attached to node C, the strongest node in this network is node A, $s_A = 1.3$. A degree-central or strength-central variable is one that can affect other variables or can be influenced by them directly without the involvement of other mediated nodes (Borgatti & Everett, 2006).

In directed networks, the degree measure of each node consists of two measurements the in-degree and the out-degree. The in-degree of node i is the total number of ingoing edges and is defined by the sum of the ith row of the adjacency matrix $k_i^{in} = \sum_j a_{ij}$. On the other hand, the out-degree of node i is the total number of outgoing edges and is determined by the sum of the ith column of the adjacency matrix $k_i^{out} = \sum_j a_{ji}$. The total number of edges of a node, called as total degree k_i^{tot} is the sum of in- and out-degrees, $k_i^{tot} = k_i^{in} + k_i^{out}$ which the total degree k_i^{tot} is equal to the degree of undirected networks ($k_i^{tot} = k_i$).

Obtaining two degrees (k_i^{in} and k_i^{out}) for each node, at least two-degree distributions are expected. The in-degree distribution, $P_{deg}^{in}\left(k^{in}\right)$ and the out-degree distribution, $P_{deg}^{out}\left(k^{out}\right)$ which are defined by the fraction of nodes in the network with in-degree and the out-degree, respectively. The third one-dimensional distribution is the total degree distribution, which $P_{deg}^{tot}\left(k^{tot}\right) = P_{deg}^{in}\left(k^{in}\right) + P_{deg}^{out}\left(k^{out}\right)$ and $P_{deg}^{tot}\left(k^{tot}\right) = P_{\deg(k)}$. For the characterization of the network structure, only the (total) degree distribution is required.

Furthermore, the strength of node i can separate into in-strength and out-strength indices. In-strength is defined by the column sum of absolute values of a node in the weighted adjacency matrix. It indicates the aggregated strength of a node that accepts the influence of other nodes of the network, $s_i^{in} = \sum_j \left|w_{ij}\right|$. Out-strength is determined by the row sum of absolute values of a node in the weighted adjacency matrix. It shows the cumulative strengths of weighted edges (w_{ij}) that outgoing from the observed node, $s_i^{out} = \sum_j |w_{ji}|$. Therefore, the summation in-strength and out-strength of node i is equal to the total strength, $s_i^{tot} = s_i^{in} + s_i^{out}$. For the weighted directed network in Fig. 4.1b, the in- and out-degrees of node B are $(k_B^{in}, k_B^{out}) = (1, 1)$ and the $k_B = k_B^{tot} = 2$; while the in- and out-strengths are $(s_B^{in}, s_B^{out}) = (0.8, 0.3)$ and the $s_B = s_B^{tot} = 1.1$.

5.2.1.2 Closeness and Betweenness

Degree and strength consider only the local structure around the observed node; however, the measure does not involve the global structure of the network. In other words, high strength does not infer that the node is central. For instance, a node might have many neighbors, but it might not be able to reach other nodes in the network quickly (Borgatti & Everett, 2006). Two more important indices could define the structure of a network depending on the distance and geodesics. The length of a shortest path among two nodes is called distance (Dijkstra, 1959) while a shortest distance path (a path with the minimum number of edges) that connect two nodes is defined as geodesics (Brandes, 2001, 2008).

In 1959, Dijkstra proposed an algorithm to identify the shortest paths (the least costly path between two nodes) in a network where the weights could be transformed into costs (Dijkstra, 1959). Later, Freeman introduced the closeness centrality of a node as the inverse farness which is defined as the sum of the distances between the observed node and all other nodes in the graph (Freeman, 1978). Years after, Newman applied Dijkstra's algorithm in weighted networks to identify the least-costly paths between all nodes (Newman, 2001). The closeness measure to be applicable in weighted networks, Newman inverted the positive weights (as Freeman accomplished) dividing one by the weight, $C_c(i) = \sum_j \left[d_{ij}^w\right]^{-1}$. Therefore, a high farness was converted into a low closeness and a low farness was transformed into a high closeness. Generally, a high closeness central node is the one that quickly reaches other nodes in the network (Wasserman & Faust, 1994) and as a result, this node communicates efficiently with all other nodes (Freeman, 1978). In Fig. 4.1a, the closeness of node C is 0.0558. The computation of this measurement is determined by the distances between node C and all other nodes in the network, A (1/0.3 + 1/0.8), B(1/0.3), D(1/0.2), E(1/0.2). The inverse sum of all distances is considered as the closeness centrality of node C.

Betweenness centrality of a node is defined as the number of times that act as a bridge along the shortest path between two nodes in binary networks (Freeman, 1978). Later, in 2005, Borgatti proposed a new and more efficient algorithm for computing betweenness on weighted networks (Borgatti & Everett, 2006). This generalization considers that the flow between two nodes can traverse faster along paths with more strongly-connected intermediate nodes than paths with fewer weakly-connected intermediate nodes. Consequently, if a node with a high betweenness centrality is removed, then the distances between nodes will generally increase.

Thus, betweenness central nodes are important in the system because they can quickly influence other nodes in the system. The betweenness can be represented as $C_B(i) = \sum_{j \neq i \neq k \in V} \frac{g^w_{jk}(i)}{g^w_{jk}}$, where $g^w_{jk} = g^w_{kj}$ denotes the total weight number of shortest paths from node j to node k (vice versa) and $g^w_{jk}(i)$ represent the weight number of those paths from j to k that pass-through node i. In the undirected network (Fig. 4.1a), the betweenness centrality of node C is 3 which is the highest. The three paths that traverse node C are those between node E and nodes A, B and D.

5.2.2 Clustering Coefficient and Transitivity

In most of the real-world networks, nodes have the tendency cluster into densely connected groups (Holland & Leinhardt, 1970). Therefore, additional network properties have been developed to evaluate this tendency and distinguish the structure of complex networks. Specifically, the local clustering coefficient evaluates the neighbors' grouping trend of a node (Watts & Strogatz, 1998) and the global clustering coefficient evaluates the overall degree of clustering in a network (Luce & Perry, 1949).

In order to define what is clustering, consider, for instance, any friendship network where nodes represent people and edges represent the friendships among nodes. In real life, two friends of a person are quite possible to be friends of each other. Therefore, local clustering Coefficient of node i, $C_{cl}(i)$, assesses the clustering of the individual node considering its neighbours. It is defined by the fraction of pairs of neighbours of a node that are also neighbors to each other (Watts & Strogatz, 1998). More precisely, a triangle is defined as a triple of nodes all connected. The clustering coefficient can be specified as the number of triangles (where the observed node is linked), normalized by the possible number of such triangles. Node C has four neighbors, A, B, D and E (Fig. 4.1a). Then, the maximum number of connections between these neighbors (MAX) equals $ki * (ki - 1)/2 = 6$. Moreover, two triangles are detected $\{A, C, B\}$ and $\{A, C, D\}$; while, the neighbors of C are neighbors of each other, $\{A, B\}$ and $\{A, D\}$. Thus, the $C_{cl}(C) = \frac{2}{6} = 0.333$.

While the clustering coefficient proposed by Watts and Strogatz can be applied only to unweighted networks, on the other hand, Barrat et al., captured the connectivity of the graph including the weights of the edges (Barrat et al., 2004). Therefore, the extended definition of clustering coefficient is the ratio of the sum of closed triplets and the total sum of weights of triplets, $C^w_{cl}(i) = 1/s_i(k_i - 1) * \sum_{j,h} a_{ij}a_{ih}a_{jh} * \left(w_{ij} + w_{ih}\right)/2$, where s_i and k_i denote respectively the strength and the degree of node i, a_{ij} indicates a binary operator of the adjacency matrix between nodes i and j and w_{ij} represents the weight of the edge among nodes i and j. For example, in Fig. 4.1a, two triangles (closed triplets) are observed $\{B, A, C\}$ and $\{C, A, D\}$ including node A; while, mutual neighbours of A are, $\{B, C\}$ and $\{C, D\}$. The weighted cluster coefficient of node A is 0.5384, $C_{cl}^{w}(A) = 1/(1.3 * 2) * \left[\frac{0.8+0.1}{2} + \frac{0.1+0.8}{2} + \frac{0.1+0.4}{2} + \frac{0.4+0.1}{2}\right]$. In general, if the clustering coefficient of node i is higher (close to 1) then most of its neighbours are also connected. This indicates that the clustering coefficient measures not only the connectivity but also the redundancy of each node (Newman, 2010). This implies that removing node i, its neighbors can be still connected or they can influence each other.

To recall, for each node, the local clustering coefficient is calculated. The local clustering coefficient measures the tendency of the nodes to be clustered in a form of triangles. On the other hand, transitivity is a property of a network and it is comprised in global network indices. Transitivity or global clustering coefficient provides an overall interpretation of the clustering in the network (Luce & Perry, 1949). This network measurement is closely related to the clustering coefficient that determines the tendency of two nodes that share a neighbor to be linked with each other, but for the entire network. Moreover, transitivity (T) is defined as three times the number of triangles in the network (the number of closed triplets), over the number of all triplets of nodes in the network (both open and closed), $T = \frac{3 \times number\ of\ triangles}{number\ of\ all\ triplets}$. The factor of three in the number of triangles indicates that each triangle contributes to three different connected triples in the graph (Newman, 2004). For instance, the number of triangles of the network in Fig. 4.1a are 2, $\{A, C, B\}$ and $\{A, C, D\}$ and 11 connected triplets. The transitivity or the global cluster coefficient of the network is 0.545 indicating that the network contains some communities of nodes that are not densely connected (within the community). The values of a local and a global coefficient fluctuate between 0 and 1 and they have closely related since both measure the relative frequency of triangles.

5.2.3 *Average Path Length*

Another global measurement is the average path length L which it is defined as the average shortest path between two nodes (Newman, 2001). Considering that d_{ij} determines the distance between node i and j; the length of a path is the number of edges that the path includes. Summarizing the above indices, another global measurement was proposed the average path length. The average path length is defined by the average steps along the shortest paths averaged over-all pairs of nodes. In an undirected graph of N nodes, $L = \frac{1}{N(N-1)} \sum_{i \neq j} d_{ij}$, where the sum is over-all pairs of district nodes. Additionally, the average path length measures the effective size of the network. In other words, this property counts how far it is on an average one node to another. For example, in a friendship network, L defines the average number of friends of the shortest path among two persons in the social network. Lastly, the average path length might not be confused with the diameter D. Both measure the size of the network, although the diameter assesses the maximal distance between all distances of any pair of nodes. The average path length of the network in Fig. 4.1a is

$L = 1.4$, which signifies that each node selecting the shortest path, requires 1.4 steps to reach to another node. Generally, the average path length is one of the three most robust measures of network topology, including clustering coefficient and degree distribution.

5.2.4 Small-Worldness

The well-known experiment in social science is the six-degree of separation by Stanley Milgram in 1967. Milgram proved that the average number of a random person to reach anyone in the social network is six steps (hops). This experiment emerged the small-world phenomenon in society and generally in real-world networks since they are neither entirely regular nor entirely random. In 1998, Watts and Strogatz proposed the small-world network model (WS model) which captures characteristics of regular (or lattice) networks and random networks. Specifically, the high clustering coefficients of regular networks and the short average path length of random networks.

They aimed to generate a social network structure to explain a phenomenon by modeling the process that creates or guides to the phenomenon. Firstly, a regular graph was used with N nodes of degree k. Then, a subset of the edges was reconnected and replaced with random edges. The number of edges that are rewired was controlled by a parameter, p'. If $p' = 0$, the graph is characterized as completely regular; on the other hand, if $p' = 1$, the graph is defined as completely random. The WS model justifies that obtains tightly interconnected cliques (clusters or groups) of nodes and few links that connect portions of the network which are far away from each other (a shortest mean path length).

Furthermore, studying dynamic systems as small-world networks, often, a generative process is used to construct an equivalent WS standard network model. By this way, the characteristics of the networks are compared and it is easier to classify them as small-world networks. Therefore, network topologies are separated into three classes of regular, random, and small-world networks (Fig. 5.1). However, the uncertainty of what degree the real-world systems in the small-world category have common network properties and to what specific point (between regular and random) were remained unsolved until 2008. Humphries and Gumey proposed a continuous, quantitative, measure of 'small-worldness', (SW) to overcome the above weaknesses. For undirected and unweighted graphs is defined by the ratio of the standardized transitivity value to the standardize average distance among two nodes (Humphries & Gumey, 2008).

Networks with small-worldness between 1 and 3 adopt part of small-world properties and they are classified as pseudo-random networks. The strictest borderline to identify a small-world network occurs when the small-worldness value of the network is higher than 3. The small-worldness can be applied not only on connected networks but also on huge segments of a disconnected network since it depends on

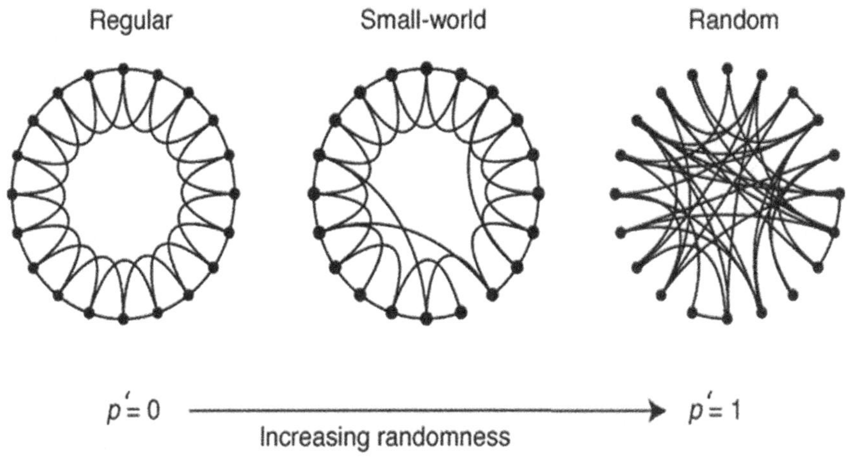

Fig. 5.1 Three basic network types in the model of WS (Watts & Strogatz, 1998). Each graph consists of 20 nodes (N = 20). At the left-side graph, each node has four neighbors (k = 4). Thus, it is defined as a high clustering coefficient C_{cl} and a high average path length L (p' = 0). Increasing the rewiring probability p indicates that randomly edges are rewired and reconnected to arbitrary nodes. Rewiring all nodes of the regular lattice network (p' = 1), the graph becomes completely random (right-side), and properties such as low clustering coefficient and low average path length are showing up. If p' value is between 0 and 1, $0 < p' < 1$, a small-world network is emerged, which incorporates the high clustering coefficient of regular networks with the low average path length of random networks

shortest paths among all pairs of nodes (Costantini et al., 2015; Humphries & Gumey, 2008).

5.3 Network Accuracy

After estimating the cross-sectional business network model, it is necessary to assess the accuracy and stability of these network structures (Efron, 1979; Epskamp et al., 2017; Epskamp & Fried, 2017). Accordingly, the '*bootnet*' R package (Epskamp & Fried, 2019) is suitable to perform bootstraps to ascertain the accuracy of edge weights and the stability of centrality indices. The bootnet function requires as input the original data and the estimation method. For the reminder, this chapter describes and illustrates the accuracy of the job-satisfaction networks and evaluates the stability of the inferences of the networks performing the case dropping subset bootstraps.

Bootstrapping (Efron, 1979) is a resampling or simulating technique that is used to estimate statistics on a population of the generating models. For ordinal data, the non-parametric bootstrap is applied, as Epskamp et al. (2017) suggest, to access the accuracy of the network. In this type of bootstrap, the observations are resampled with replacement to create new, reasonable datasets. Moreover, some observations might be sampled more than one time; while, other observations might by no means be sampled. Therefore, this process of sampling with replacement is similar to the true model due to repeatedly sampling from the population. The non-parametric bootstrap is used to determine confidence intervals (CIs) around the regularized edge-weights (Hastie et al., 2015).

For example, 1500 random samples are creating from the original dataset. Afterwards, the mean of the bootstrap sample is calculated. This process is repeated in order to generate a bootstrap sampling distribution of the means. For the sake of simplicity, the means of the 95% of the resulted CIs of the bootstrap samples are visualized. The area outside of the 95% CI indicates that the probability of the true mean to fall in this area occurs less than 5%. In other words, the p-value is less than 0.05 signifying that there is a statistically significant difference between the true mean and any value outside the CI. Another usage of a visual statistical test could appear when two or more samples are compared. If the 95% CI does not overlap, it is inferred that there is a statistically significant difference in the weights of the samples. On the other hand, if the 95% CI of the samples overlaps, then the means are compared if they are significantly different from each other.

In conclusion, many factors can affect the accuracy and stability of results such as the sample size and the true network structure (Epskamp & Fried, 2017). In general, smaller sample sizes estimate empty networks or networks with very few edges. On the contrary, complete or very dense networks, including partial correlations above 0.8 indicate near-perfect collinearity among variables. For the record, the covariance matrix's estimators fail to be positive definite and the matrix is called singular (Chtelat & Wells, 2016). Mainly, the singularity is caused by collinearity which it is caused when two variables are highly correlated to each other (Nisbet et al., 2018) and it can be inspected through a simple correlation matrix among all variables. Both cases lead to difficulties in interpreting networks. In addition, simulations have been shown that the bootstrap samples have been significantly different from the true sample (Epskamp et al., 2017; Epskamp & Fried, 2017).

5.4 Network Centrality Stability

In order to evaluate the stability of the inferences of the network, it is suggested to perform the *case-dropping subset bootstrap* (Epskamp et al., 2017). Moreover, this bootstrap is known as *m out of p bootstrap*, which *m* indicates the observation (rows); while *p* points out the nodes or variables (columns of the dataset). The term stability inspects how the order of the centrality indices remains steady, after the re-estimation of networks using fewer observations from the initial dataset (subsampling without replacement).

Furthermore, the bootstrap method minimizes the sample size from 95% to 5% of the initial dataset and the quantification of centrality stability is accomplished with the Correlation Stability-coefficient (CS-coefficient). The CS-coefficient measures the 95% of the confidence intervals of the correlations that are equal or higher than

0.7 (*CS*(*cor* = 0.7)) between the centrality indices of the original network and the centrality indices of the total subsampling datasets. If these correlations are modified significantly after excluding only 10% of the cases, the stability will be decreased because different weights will lead to different centrality indices and, consequently various networks will be generated. Ideally, the CS-coefficient could be above 0.5, $CS(cor = 0.7) \geq 0.5$, but not lower than 0.25 (Epskamp et al., 2017).

Even though a network is considered stable, it could not be interpretable due to the results of the network structure. Specifically, the centrality indices do not present any modification. For instance, all nodes have the same betweenness score (Epskamp & Fried, 2017).

5.5 Network Analysis in Business

Network theory is suitable for inferring networks related to our job satisfaction problem. Once the network is estimated, different indices can deduce the structure and examine the relative importance of the nodes (factors). Specifically, it is necessary to notice which of the ten job satisfaction factors can be the most significant in order to distinguish how they can influence directly other nodes (strength centrality). This could be useful for companies to make decisions, e.g. keeping their personnel satisfied by improving only a minimum number of features/nodes. Furthermore, it is possible to recognize which factors are instantly affected (directly or indirectly) by various modifications in any part of the system (closeness centrality). Equally important is the job attractiveness factor that is capable of control and flow faster the information between all pairs of nodes in the graph (betweenness centrality). Finally, it is necessary to investigate if a graph tends to cluster together. The topology of the graph will provide a wider knowledge of its properties concerning the generation and the evolution of the network (Barabási & Bonabeau, 2003). For example, if a new factor is added in the job-attractiveness network, which of the current nodes could be connected (small-worldness index)? Additionally, it is possible to inspect the relative importance of a job-satisfaction factor when its removal changes the stability of the system. This signifies that the information will not flow normally in the network since the neighbors of the observed node are not connected (local clustering coefficient).

The data-driven network was created using the *bootnet* package in R (Epskamp & Fried, 2019; Epskamp et al., 2017), as Chap. 4 describes. Below, the most notable measurements such as centrality, shortest path length, clustering and small-worldness are applied to the weight matrix of each working sector. These indices will summarize the patterns of connections in the network to identify which of the ten factors plays a significant role in the job attractiveness problem.

5.5.1 Centrality Analysis in a Job-Attractiveness Problem

As it was previously noted, the centrality of a node is considered from three indices, the strength, closeness and betweenness. Various centrality measurements can be easily computed with the function *centrality_auto* from the qgraph package (Epskamp et al., 2012). In particular, the required input of the centrality function is the weight adjacency matrix derived from the graphical lasso method (Chap. 4). The indices of degree and centrality (betweenness, closeness and strength) of the Retail sector are presented in Fig. 5.2b. Considering that human reasoning cannot effectively handle numerical values, Fig. 5.2c, d visualize the degree distribution of

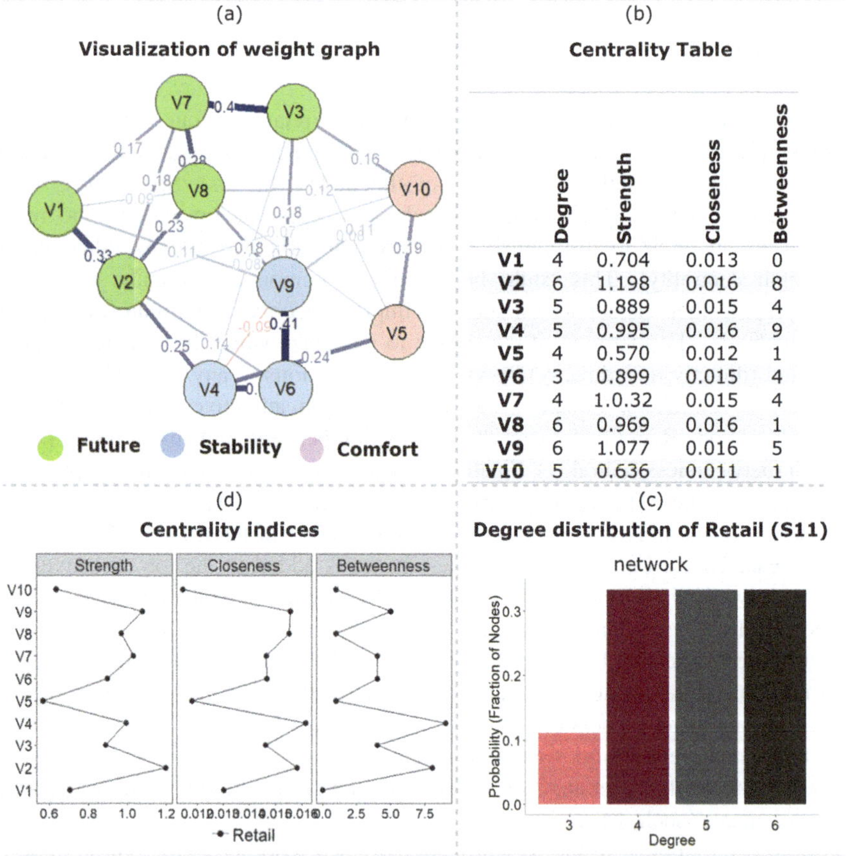

	Degree	Strength	Closeness	Betweenness
V1	4	0.704	0.013	0
V2	6	1.198	0.016	8
V3	5	0.889	0.015	4
V4	5	0.995	0.016	9
V5	4	0.570	0.012	1
V6	3	0.899	0.015	4
V7	4	1.0.32	0.015	4
V8	6	0.969	0.016	1
V9	6	1.077	0.016	5
V10	5	0.636	0.011	1

Fig. 5.2 Representation of the job satisfaction graph of Retail section (top left). The color of the nodes denotes the classification in one of the three clusters; Future (green), Stability (blue) and Comfort (pink). Figure 5.2b reports the values of degree, strength, closeness and betweenness indices of ten nodes (top right). The degree distribution shows the spread of nodes' degrees in the network (bottom right) and the visualization of the raw centrality measures for each node is demonstrated at the bottom left

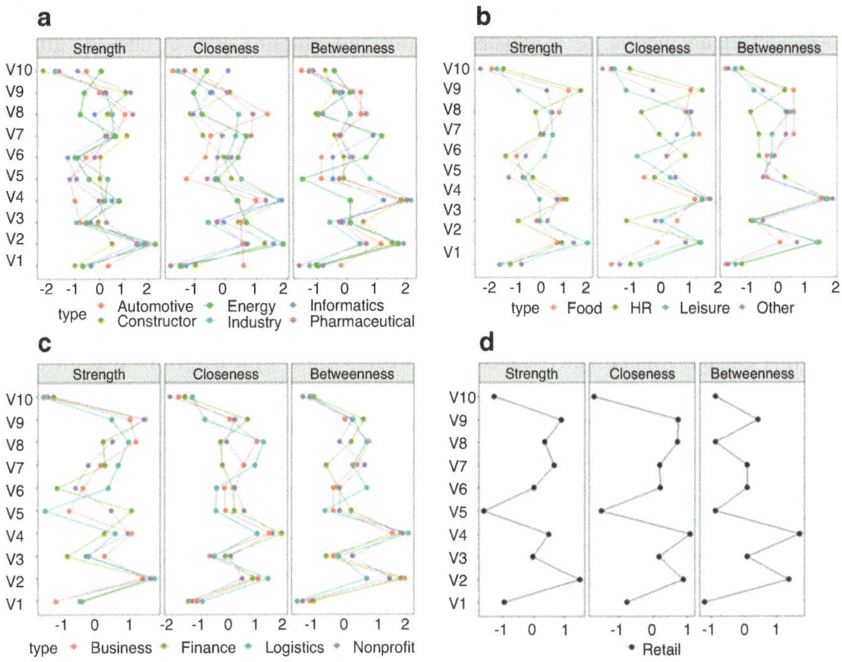

Fig. 5.3 The importance of each node of the job-attractiveness networks quantified in centrality indices. Centrality indices (strength, closeness and betweenness) have been standardized (z-scores) to facilitate the comparisons between the job satisfaction networks

the job satisfaction network and the centrality coefficients estimates, respectively. Furthermore, Fig. 5.3 shows the standardized centrality indices of the 15 sectors which are classified into four groups according to the categorization that is obtained from *k*-means clustering method (Chap. 4). For instance, the first group, (Future: V_1, V_2, V_7, V_8; Stability: V_4, V_6, V_9; Comfort: V_3, V_5, V_{10}) includes the following six working sectors: Automotive, Construction, Energy, Industry, Informatics and Pharmaceutical (Fig. 5.3a). The fourth group contains only one, the Retail sector, as it is presented in Fig. 5.3d and Fig. 5.5d. The centrality-indices of Fig. 5.3 have been standardized (z-score) in order to be on the same scale. This could facilitate the comparisons between the job-attractiveness networks.

It is reported that some nodes are consistently more central or peripheral in job-attractiveness networks (Epskamp et al., 2017). Particularly, the majority of sectors mention that the centrality measurements indicate that factors long-term job security (V_4) and Prospects/career opportunities (V_2) are the most central nodes in the job-attractiveness networks. Long-term job security (V_4) is noticed as the betweenness-central and closeness-central node, as well as Prospects/career opportunities (V_2) is determined as the highest strength-central node. Overall, the most significant nodes lie in clusters Future and Stability for 'a' and 'd' groups (Fig. 5.3a, d), solely in cluster Future for group 'b' (Fig. 5.3b) and Future and Comfortable clusters for group 'c' (Fig. 5.3c).

Notably, V_4 is considered as the betweenness-centrality factor in most of the working sectors; while, Informatics (S_7), Energy services (S_{10}) and Business services (S_{12}) indicate the Prospects/career opportunities (V_2) as the first betweenness-central factor (Fig. 5.3). This specifies that V_4 owes its high betweenness to the ability to control the flow between all pairs of nodes in the network, under the assumption that information mainly travels through the shortest paths between them. As an illustration, in the Retail sector (Fig. 5.2a), the information traverse Future and Stability clusters. Specifically, the edge between V_4 and V_2 is determined as the most betweenness-central edge, since most of the shortest paths among V_2 and other job-attractiveness features traverse the edge V_2–V_4 and then, V_4 propagates the influence to other job-attractiveness features. In other words, if we decrease the importance of V_4, the influence will hardly flow over the rest job-satisfaction features and vice versa. Moreover, V_2 is determined as the second betweenness-centrality factor, excluding six sectors. Finance (S_3), Food (S_4) and Non-profit (S_9) point out the factor Strong management (V_9); Industry (S_6) indicates the factor Good balance between private work and life (V_5); Energy (S_{10}) denotes factor Interesting jobs (V_7) and Informatics (S_7) points to V_4.

In most sectors, V_4 (Long-term job security) is considered as the closeness-centrality node due to very short paths that connect it to other trails. This implies that a change in a part of job-attractiveness networks will instantly modify the reliability of the job-security factor. On the other hand, if a job provides additional security to an employee, important changes will anticipate in the entire network. Energy services (S_{10}) and Transportation (S_{13}) infer a factor that belongs in Future dimension, the Prospects/career opportunities (V_2). Moreover, Automotive (S_1) considers that Quality of training (V_8) as the most important factor; while, Food (S_4) points to V_7. Various factors are observed in the second position in which 66.7% of the sectors indicate variables that belong to Future cluster. In particular, 46.7% of the observed variables indicate Prospects/career opportunities (V_2) as the second most important factor that spread the information efficiently through a graph. Next, Quality of training (V_8) represents 20% of the sectors, such as Pharmaceutical (S_9), Transportation (S_{13}) and Others (S_{15}). However, Automotive (S_1) and Food (S_4), point out the factor V_4; Human Resources (S_5) select V_9, Non-profit (S_8) infer V_5 and Energy Services (S_{10}) determine Pleasant working atmosphere (V_3) in the second place.

Figure 5.3 points out a feature from Future cluster, Prospects/career opportunities (V_2), as a strength-centrality factor of 80% of the job-attractiveness networks; except for Food (S_4) and HR (S_5) sector which determine factor V_9 and Construction (S_2) sector that indicate factor V_7. In this certain case, V_2 can affect directly, strongly and weakly, a considerable number of nodes in the network and vice versa. Notably, for all job-attractiveness networks, V_2 highly and straight influences not only the factors that constitute the Future cluster viz., Competitive package (V_1), Interesting jobs (V_7) and Quality of training (V_8), but also V_4 that inferred as the betweenness-central and closeness-central factor to the majority of the networks. Furthermore, the focal factor V_2 is connected to Pleasant working atmosphere (V_6) and Strong management (V_9) signifying that Future cluster (specifically V_2) can directly influence Stability rather than Comfort and in reverse. The edge among V_2 and V_3 (Future—Comfort) is appearing in 13 sectors, where the strength of this link is observed as moderate in five sectors. Finally, a weak connection is observed among V_2 and Good balance between private life and work (V_5) in 40% of the job-attractiveness networks. Five sectors (S_2, S_3, S_7, S_8 and S_{11}) converge in indicating that node Strong management (V_9) is the second strength-central node. Moreover, V_4 occurs in four sectors (S_4, S_5, S_4 and S_5), four sectors (S_1, S_9, S_{12} and S_{13}) point out Quality of training (V_8) and finally, the factor Interesting jobs (V_7) is noticed in Industry (S_6) and Travel (S_{14}).

In contrast, the peripheral nodes (the nodes that cannot influence neither direct nor indirect other nodes in the network) are the Competitive package (V_1) and Deliberately handles the environment and society (V_{10}). These two factors sorted last in centrality ranking in most of the job-attractiveness networks; except for the Automotive sector which V_1 is ranked in the fifth position for betweenness and third position for the closeness-centrality. Furthermore, in Automotive and Energy Services, the strength-centrality of V_1 is ordered in the fourth place and in Finance, Industry and Informatics is ranked in the seventh place. Finally, in Informatics, V_{10} is ordered third in the closeness-centrality ranking. The strength index positioned V_{10} in the fourth and sixth place in the Energy Services and the Automotive sectors, respectively.

Table 5.1 reports a Pearson product-moment correlation coefficient (Pearson's correlation, for short; Pearson, 1900) to determine the relationship between three indices of node centrality together with the clustering coefficient of each working sector. The centrality correlations indicate the converged central node according to three centrality indices. There are positive correlations between all measurements of centrality. Specifically, positive significant correlations are observed between betweenness and closeness estimates ($.724 \leq r \leq .965$, $n = 10$, $.018 \geq p \geq .000$). Moreover, there are many strong, positive correlations between centrality indices. The weaker correlation ($p \geq .05$) between all indices of centrality are detected in Automotive (S_1) and Construction (S_2). Particularly, S_2 obtains a small positive correlation which is not statistically significant ($r = .279$, $n = 10$, $p = .435$), suggesting that the proportion of the variance that is shared by strength and betweenness is only 7%. To put it in another way, the rankings of strength and betweenness are significantly different.

On the contrary, the stronger significant correlations ($p < .01$) are detected in Food (S_4), Business Services (S_{12}) and Travel (S_{14}) sectors. Finally, according to K-mean (Fig. 4.6), the group of sectors that present high, significant correlations of all centrality indices is group 'b' and 'd'. (As a note, the correlations between

Table 5.1 Pearson's correlations of node centralities and clustering coefficients of 15 sectors. Strength is correlated strongly and positively with Closeness and with Betweenness Centrality; likewise, the association between Betweenness and Closeness is positive and high. On the contrary, the Clustering coefficient is negatively associated with centrality measurements in the majority of the sectors

	Strength		Betweenness	Clustering Coefficient		
	Betweenness	Closeness	Closeness	Betweenness	Closeness	Strength
S1	.348	.420	.812**	-.213	.218	.064
S2	.279	.343	.965**	-.424	-.503	-.747*
S3	.690*	.561	.891**	-.589	-.444	-.778**
S4	.890**	.846**	.881**	-.728*	-.654*	-.876**
S5	.781**	.757*	.839**	-.130	-.476	-.115
S6	.503	.647*	.954**	-.116	-.000	-.197
S7	.754*	.501	.811**	-.010	.134	-.527
S8	.780**	.733*	.889**	.038	.161	.029
S9	.550	.712*	.894**	-.110	-.044	.018
S10	.656*	.720*	.785**	-.524	-.271	-.477
S11	.753*	.922**	.742*	-.539	-.454	-.521
S12	.874**	.855**	.929**	-.536	-.623	-.784**
S13	.698*	.776**	.724*	-.294	.316	-.126
S14	.779**	.766**	.830**	-.440	-.020	-.456
S15	.745*	.733*	.940**	-.129	-.053	-.350

**. Correlation is significant at the 0.01 level (2-tailed).
*. Correlation is significant at the 0.05 level (2-tailed).

*Correlation is significant at the 0.05 level (2-tailed)
**Correlation is significant at the 0.01 level (2-tailed)

centrality indices and clustering coefficients will be described in the following sub-chapter).

5.5.2 *Cluster Coefficients in a Job-Attractiveness Problem*

Many centrality clustering coefficients can be computed and visualized using functions that implemented in qgraph package. In the following analyses, Barrat weighted clustering coefficient method (Barrat et al., 2004) is applied to distinguish the redundancy of a node in its neighborhood.

Clustering coefficients of the majority of the sectors converge in pointing out that Financially sound (V_6) is the most locally redundant node (Fig. 5.4). The reason that V_6 obtains high clustering coefficient depends on the tendency to be strongly connectivity of its neighbors. Therefore, V_6 is considered as a peripheral node in many job-satisfaction networks (46.7%), apart from Automotive, Industry, Energy Services and Transportation that indicate Quality of training (V_8). Moreover,

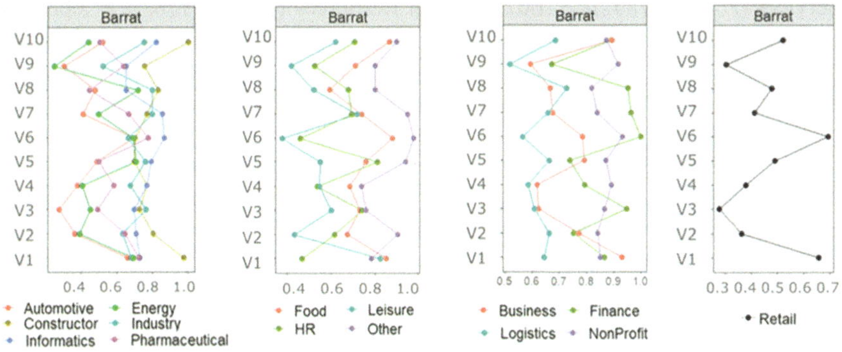

Fig. 5.4 The clustering coefficient of each node of 15 job-attractiveness networks which are separated into four groups according to K-means. As the local clustering coefficient is closer to one, indicates the redundancy of the node in its neighborhood

Business (S_{12}) and Travel (S_{14}) sectors consider Competitive package (V_1) as the highest clustering coefficient variable, Construction (S_2) points out the variable Dealing with the environment and society (V_{10}) and HR (S_5) refers Good balance between private life & work (V_5).

Thus, the second higher clustering coefficient node of four sectors is determined Interesting jobs (V_7). Each of the following variables, Competitive package (V_1), Pleasant working atmosphere (V_3) and Good balance between private life and work (V_5), points out three sectors. Strong management (V_9) and Dealing with the environment and society (V_{10}) indicate Non-Profit (S_8) and Food (S_{10}), respectively. On the contrary, Strong management (V_9) and Prospects/career opportunities (V_2) present the lowest clustering coefficient to all job-attractiveness networks.

5.5.3 *Associating Clustering Coefficients and Centrality*

In network analysis, it is useful to investigate if there is any correlation between centrality indices and clustering coefficients. Table 5.1 reports the Pearson correlations between three centrality indices and clustering coefficients in 15 working sectors. As it is observed, clustering coefficients correlate significantly negative (or weakly positive) with all centrality measurements. Specifically, there are strong, negative correlations which are statistically significant between clustering coefficients and betweenness-centrality ($r = -.728$, $p = .017$), between clustering and closeness ($r = -.654$, $p = .040$) and among clustering and strength ($r = -.876$, $p = .001$) in Food (S_4). Additionally, more statistically significant negative correlations between clustering coefficient and strength-centrality are perceived in Business

Services ($r = -.784, p = .007$), Human Resources ($r = -.778, p = .008$) and Construction ($r = -.747, p = .013$). According to the above, it is inferred that the most central nodes (V_2 and V_4) do not obtain high clustering coefficients. To justify this, a deeper investigation of associations between centrality and clustering coefficients is presented (Table 5.1).

Figure 5.5 visualizes the plots of centrality and clustering coefficient of each node of the Retail working sector (S_{10}). A vertical line notes the median value of the clustering coefficient and a horizontal line indicates the median value of the focal centrality index. Consequently, in the Retail Sector (Table 5.1), Financially sound (V_6) is the redundant node of the graph, since it obtains the highest clustering coefficient, $C_{cl}(V_6) = 0.693$.

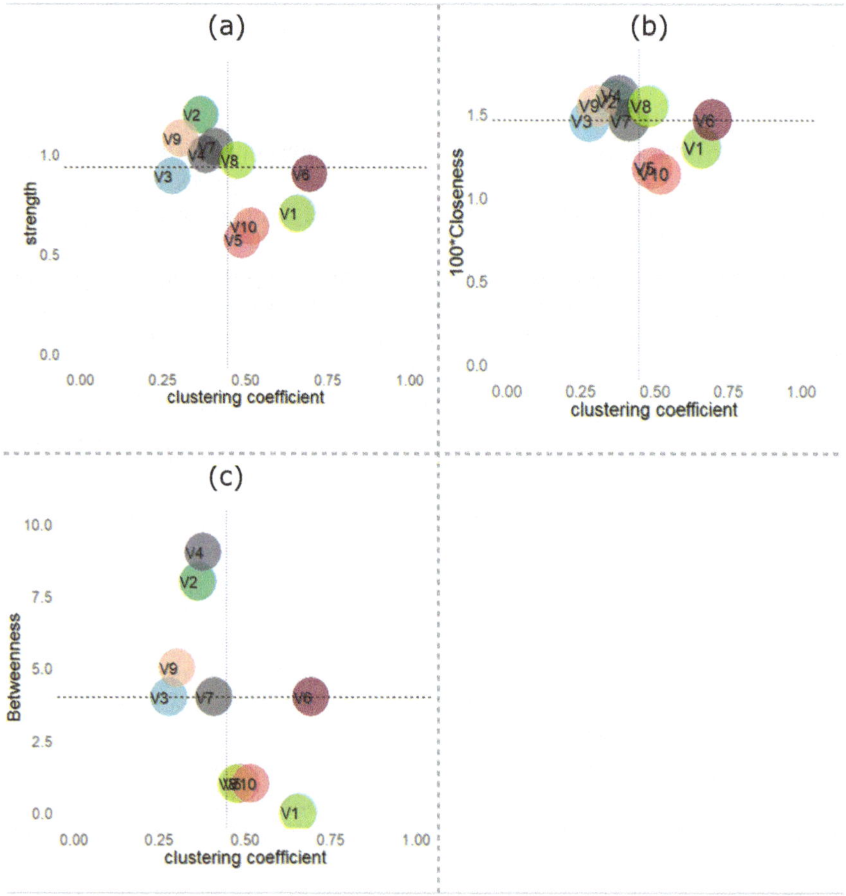

Fig. 5.5 Combining centrality measurements and clustering coefficients retail working sector (S_{10}). The most central nodes (V_2, V_4 and V_9) do not present high clustering coefficients. The values of closeness are multiplied by 100

The second highest clustering coefficient factor is Competitive package (V_1), $C_{cl}(V_1) = 0.656$, and then Dealing with the environment and society, $C_{cl}(V_{10}) = 0.519$. Due to their high clustering coefficients and low centrality indices, the nodes V_1 and V_{10} (less of V_6 because the centrality indices are considering as moderate) occupy peripheral positions in the Retail's job-attractiveness network (Fig. 5.2a). This signifies that if a high clustering coefficient node is excluded from the network, its neighbors could be connected directly to each other or using the shortest path.

In the case of V_6, the removal of its edges will not cause a significant change in the model. The information will travel through its neighbors (V_2, V_4 and V_9) since these nodes are connected straight to each other and indirectly between V_2 and V_9 through V_4. In general, the higher is the clustering coefficient of a node, the more direct connections between its neighbors ($C_{cl} > 0.75$). Furthermore, Fig. 5.5 demonstrates the nodes V_1 and V_{10} occupy places below the horizontal line (smaller values comparing to median centrality indices) since they acquire lower centrality indices comparing to V_6. Hence, V_6 is observed as a node with moderate clustering coefficient and moderate centrality indices.

On the contrary, nodes with high centrality such as V_2, V_4 and V_9 placed above horizontal line (moderate to high centrality) and at the left side of the vertical line (low clustering coefficients). The conclusion of the above investigation infers that there is an inversely (moderate to low, $p \leq .05$) proportional relationship between centrality indices and clustering coefficients for the focal job-attractiveness network.

5.5.4 Small-Worldness in Job-Attractiveness Networks

It is well known that random ER (Erdős-Rényi) and small-world networks present homogeneous degree distributions indicating that the nodes have approximately the same number of edges (Erdös & Rényi, 1959; Watts & Strogatz, 1998). In a weighted network, it is appropriate to inspect the normality distribution of the strength centrality (Costantini & Perugini, 2016). For this reason, the Shapiro-Wilk test (Shapiro & Wilk, 1965) is applied to the strength centralities of each job-attractiveness network because it is suitable for small sample sizes (<50 samples).

It is concluded that strength-centrality distribution of every sector is not significantly different from the normal distribution, as the Sig. value under the Shapiro-Wilk column is greater than 0.05 (Table 5.2). The only exception is observed in Energy Services (S_{10}), where the p-value is less than 0.05, ($Shapiro - Wilk = .838$, $p = .042$). It such case, the degree distribution is examined.

Figure 5.6 displays two plots to determine graphically the normality of the S_{10}. The Normal Q-Q Plot (Fig. 5.6a) demonstrates the data points that lay close to the diagonal line, apart from one point that strays away from the diagonal line which confirms the numerical value of Shapiro-Wilk test ($p = .05$). Moreover, the degree distribution follows a right-skewed distribution ($skeweness = 1.487$), signifying that

Table 5.2 Descriptive statistics and test of normality of (N = 10) strength-centrality values of 15 sectors. Observing the p-values of the Shapiro-Wilk test, the strength of every sector follows a normal distribution, except for S_{10} ($p < 0.05$)

	Descriptive statistics					Tests of normality Shapiro-Wilk	
	Mean	SD	Median	Skewness	Kurtosis	Statistic	Sig.
S1	0.926	0.090	0.903	0.715	−0.249	0.942	0.580
S2	0.935	0.119	0.949	−1.381	2.398	0.867	0.091
S3	1.044	0.170	1.064	0.119	−1.322	0.927	0.422
S4	0.975	0.192	1.004	−0.436	−1.213	0.929	0.437
S5	1.006	0.205	0.997	0.403	−1.125	0.930	0.443
S6	0.962	0.132	0.996	−0.115	0.409	0.962	0.808
S7	0.997	0.149	0.958	0.708	−0.220	0.943	0.592
S8	1.051	0.178	1.051	−0.031	−1.155	0.955	0.733
S9	0.917	0.160	0.931	−0.048	−0.335	0.955	0.731
S10	0.949	0.193	0.860	1.478	1.996	0.838	0.042
S11	0.897	0.202	0.934	−0.387	−0.804	0.950	0.665
S12	0.933	0.164	0.945	−0.305	−1.457	0.918	0.344
S13	0.940	0.138	0.981	−0.458	−0.262	0.933	0.482
S14	0.876	0.194	0.908	0.267	0.312	0.950	0.673
S15	0.956	0.130	1.001	−0.851	0.706	0.943	0.586

many nodes have few connections (Fig. 5.6b). On the other hand, Fig. 5.6c, d present the Normal Q-Q Plot of S_6 that gathers the highest value of Shapiro-Wilk test of strength-centrality (the data points are close to the diagonal line) and the degree distribution, respectively.

Regarding the normal distribution of strength-centralities, we hypothesize that our networks show a small-world property. For this reason, the homonym function of qgraph is used to calculate the small-worldness index. First, the function converts the weighted graph to an unweighted graph (it is also called binary graph) to compute the average degree distribution ($\overline{k}$), the average path length ($\overline{L}$) and the transitivity (T) of the network. Then, 1500 arbitrary networks are generating using as input the degree distribution of the focal network and the number of the observed nodes. Thus, the random seed was set equal to one to ensure the replicability of the results of the estimated networks. The function returns the values of the estimated transitivity ($\boldsymbol{T_{est}}$) and average path length ($\boldsymbol{L_{est}}$) of the estimated random networks to make comparisons with the actual indices of the focal network. Finally, the small-worldness (SW) index is computed.

Table 5.3 reports the results of small-worldness index of each working network, the values of average path length ($\overline{L}$) and transitivity (T) of the actual and estimated networks (L_{est} and T_{est}). The results show that there are no significant differences between the indices of average path length and transitivity of actual networks and those that were estimated from similar networks. Furthermore, the small-worldness values of the job-attractiveness networks fluctuate between 0.951 and 1.141. These

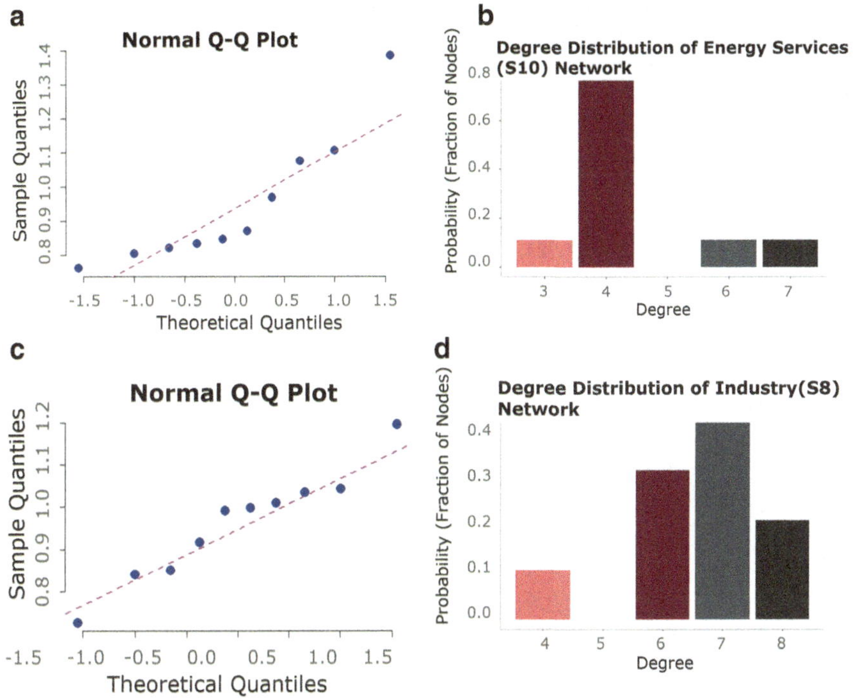

Fig. 5.6 Normal Q-Q plots and degree distributions of the least normal distributed strength-centrality (**a** and **b**) of the S_{10} dataset and the highest normal distributed strength-centrality (**c** and **d**) of the S_6 dataset

values are lower than 3, indicating that none of the networks holds the small-world property. In contrast, in a completely random graph consisting of N nodes, the average clustering coefficient is much smaller than one, $\overline{C}_{cl} \sim 1/N$. Consequently, our job-attractiveness networks are characterized as pseudo-random networks since they appear modest SW properties, i.e., high clustering coefficients that are even higher comparing to a regular ring lattice, $(\overline{C}_{cl} \cong 0.75, p' = 0)$; while, maintaining the average path length as low as possible, compared to a random graph (Erdös & Rényi, 1959; Watts & Strogatz, 1998).

The results of small-worldness index infer that the nodes of each network do not indicate the tendency to be highly clustered. Moreover, there is no possible prediction about the connection probability of a new node in the network, since the important properties of random graphs can appear quite suddenly (Erdös & Rényi, 1959). Furthermore, it is inferred that the connection probability of each node of the observed job-attractiveness networks is approximately 0.1 $(p \simeq 0.1)$ due to homogeneous degree distributions. This denotes that large hubs do not emerge in the

Table 5.3 According to small-worldness index of each working network, none of the networks holds the small-world property since $SW < 3$. Moreover, the values of Transitivity (T) and Average Path Length ($\overline{L}$) are similar to the estimated T_{est} and L_{est} of random graphs. Consequently, these job-attractiveness networks are classified as pseudo-random

	Average degree $\overline{k}$	Average clustering $\overline{C}_{cl}$	Transitivity T	Average path length $\overline{L}$	Small-worldness SW	Estimated T_{est}	Estimated L_{est}
S1	5.6	.507	.522	1.378	.950	.550	1.378
S2	7	.803	.784	1.222	1.003	.782	1.222
S3	7.2	.858	.837	1.200	1.019	.820	1.200
S4	6.4	.749	.730	1.289	1.040	.701	1.289
S5	6	.624	.646	1.333	.986	.655	1.333
S6	6.6	.706	.707	1.267	.986	.717	1.267
S7	6.8	.738	.735	1.244	.986	.746	1.244
S8	8	.870	.865	1.111	1.000	.865	1.111
S9	5.8	.597	.596	1.356	.973	.615	1.356
S10	4.4	.526	.481	1.511	1.141	.426	1.527
S11	4.8	.457	.438	1.467	.951	.463	1.475
S12	6.4	.734	.713	1.289	1.014	.703	1.289
S13	6.4	.632	.648	1.289	0.981	.660	1.289
S14	5.4	.562	.532	1.400	0.972	.549	1.403
S15	7.6	.856	.852	1.156	1.007	.846	1.156

network. Consequently, every new node can attach every single node of the network with the same probability (random graph property).

5.6 Network Accuracy in Business Networks

In these job-satisfaction structures, the weight of the edge between two nodes is a parameter estimated from the data. As the sample size increases more accurate the learning parameters are (Epskamp et al., 2017). This indicates that the estimated networks derived from very large datasets are reflecting the true model. Nevertheless, in every sector, the number of observations is not always sufficient. Consequently, many spurious edges will be estimated and the interpretation of the network will be uncertain. Therefore, as Epskamp et al. (Epskamp et al., 2017; Epskamp & Fried, 2017) advice in estimating psychological networks, a nonparametric bootstrap method is applicable for the business networks to assess the accuracy of the network parameters.

Next, the bootnet (Epskamp & Fried, 2019) R package is utilized to perform the nonparametric bootstrap (sampling with replacement). In particular, it is possible to resample a single observation more than once, while others may not be resampled at all. The repeated samples from the population will be similar to our initial model. Then, the mean of the bootstrap samples is calculated and will be compared with the actual weights. To run bootstraps, the function uses the same arguments for estimating the network. Moreover, all simulations estimate partial correlation matrices of ten nodes using the graphical lasso method in combination with the EBIC model selection (Epskamp et al., 2017; Epskamp & Fried, 2017; Costantini et al., 2015; Foygel & Drton, 2010). In this study, we set the number of samples to 1500 for every sector.

Figures 5.7 and 5.8 illustrate the resulting plots of 15 job-satisfaction networks of different sample sizes. The edges are ordered according to their strength. Further, the red line indicates the sample (true) values of the network's weights and the black line represents the mean of the weights of the 1500 bootstraps. The grey area denotes the bootstrapped CIs around the regularized edge weights of the network. To recall, CIs

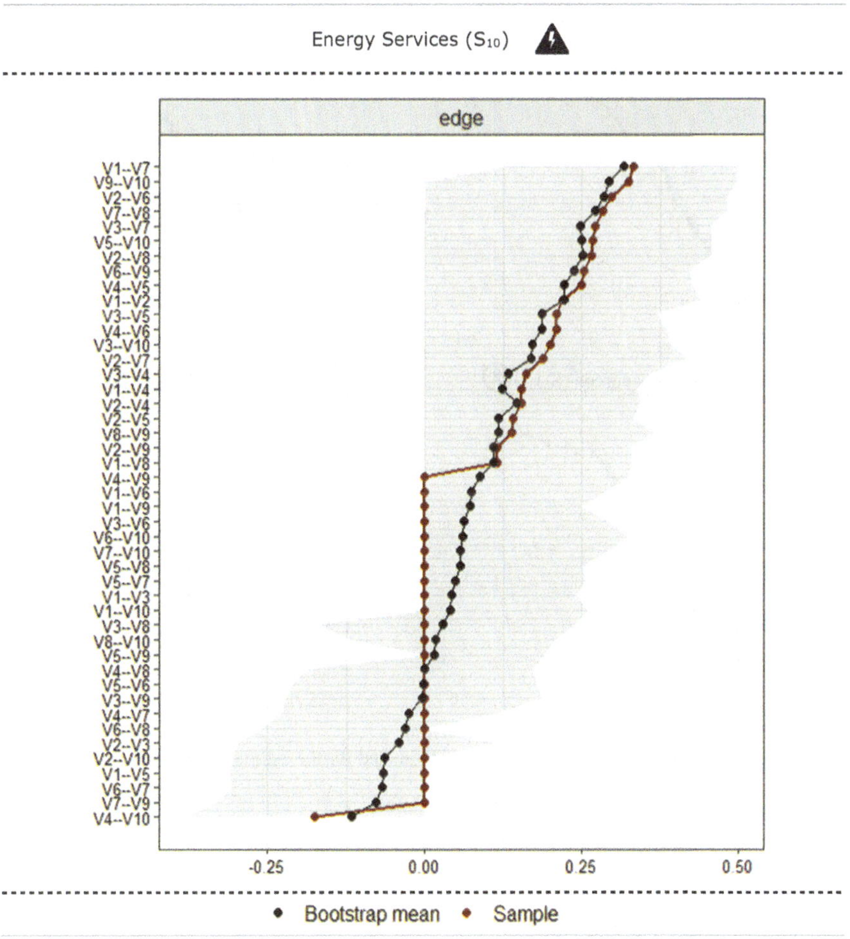

Fig. 5.7 Bootstrapped confidence intervals for the estimated job-satisfaction network in the Energy sector

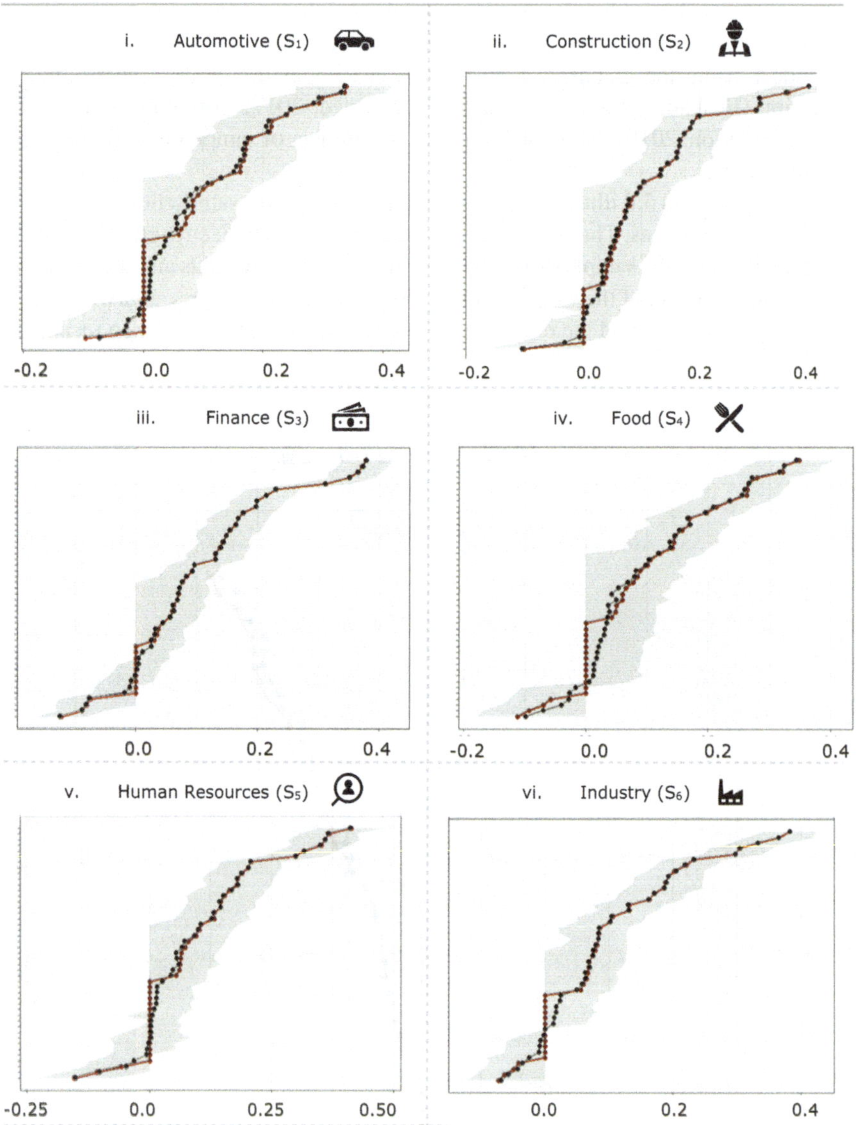

Fig. 5.8 Bootstrapped confidence intervals of estimated for the estimated job-satisfaction network in 14 working sectors

represent 95% of the bootstrap samples indicating that the probability of the value to be in this area is 0.95. To put it differently, the values near the edge of the distribution tend to be rare compare to values that are closer to the mean of the bootstraps. The y-axis depicts the pairs of variables, represented by horizontal lines and ordered according to the strength of edge weight (from positive to negative). For instance, the highest positive weight in the Energy sector (Fig. 5.7) is observed

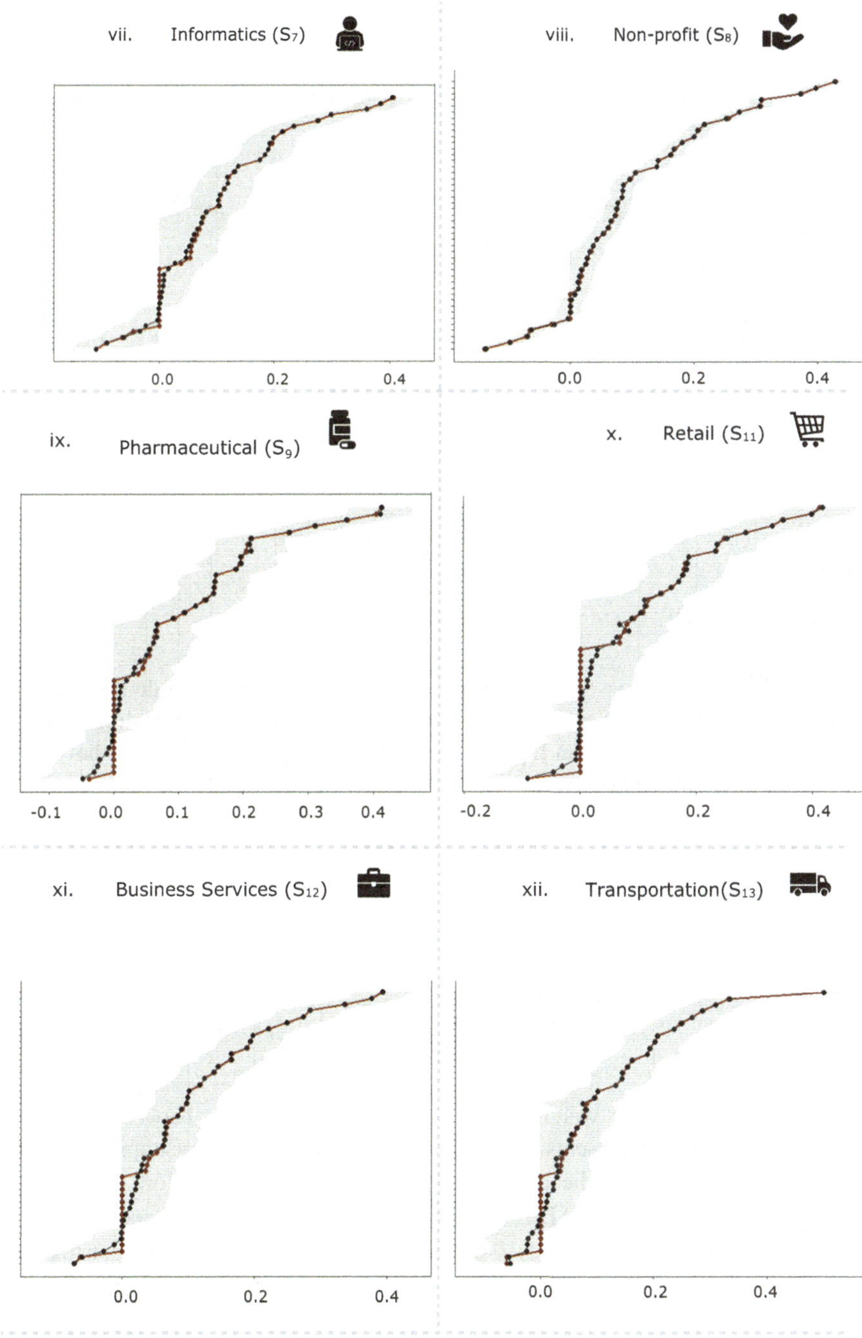

Fig. 5.8 (continued)

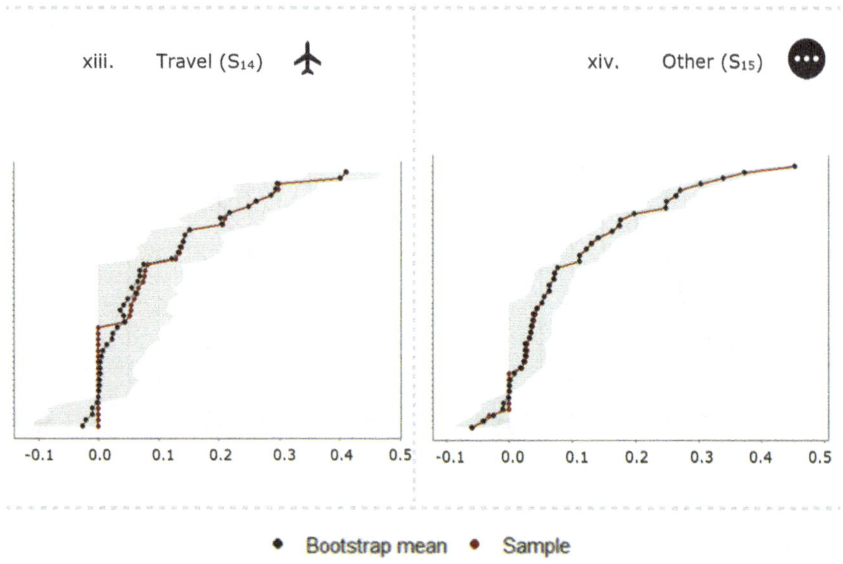

Fig. 5.8 (continued)

among V_1 (Competitive salary package) – V_7 (Interesting job), while the strongest negative weight is between V_4 (Long-term job security) – V_{10} (Dealing with the environment and society). In Fig. 5.8, the labels of the y-axis of the plots have been removed to avoid derangement.

Moreover, the plots in Figs. 5.7 and 5.8 reveal the differences in each pair of variables between the sample and the mean bootstrap values. In an effort to quantify the differences of the red and black lines, we calculate the Sum of the Absolute Differences (SAD) among the sample weights and the mean bootstrap weights (Table 5.4):

$$SAD = \sum_{i=1}^{n} \sum\nolimits_{j=1}^{n} \left| w_{ij} - \overline{w}_{boot\,ij} \right| \tag{5.1}$$

where w_{ij} denotes the weight of the sample (true) network between the i and j variables, $\overline{w}_{boot\,ij}$ represents the mean weight of the 1500 bootstraps among the variables i and j and n indicate the number of the variables. Therefore, in every job satisfaction network, the SAD is calculated with 45 ($\frac{n(n-1)}{2}$) parameters.

Additionally, Figs. 5.9 and 5.10 depict the quantile intervals of the bootstrapping method only for the times the parameter (the weight between two nodes) was not estimated as zero. Furthermore, the values inside the boxes in the plot indicate the percentage of the parameter to be zero. Moreover, the transparency of the intervals denotes the frequency of an estimated edge in the network. Finally, the pair of nodes

Table 5.4 The sum of the absolute differences (SAD) among the sample weights and the mean bootstrap weights. Lower SAD values are observed in larger datasets (S_8 and S_{15}) indicating more accurate networks since the differences between the true model and the mean of the 1500 bootstrap models is very low ($SAD \leq 1$) while higher SAD values are detected in smaller datasets (such as in Energy sector, S_{10})

Sectors	$\sum_{i=1}^{n}\sum_{j=1}^{n}\left\|w_{ij}-\overline{w}_{boot_{ij}}\right\|$	Rank	Number of observations
S_8	0.0622	1	20,378
S_{15}	0.0763	2	7667
S_3	0.1406	3	3819
S_7	0.1556	4	3262
S_2	0.1889	5	2301
S_{12}	0.2031	6	3134
S_6	0.2327	7	3053
S_9	0.2623	8	2363
S_{13}	0.2738	9	2833
S_5	0.2885	10	1499
S_{11}	0.3153	11	1469
S_{14}	0.3528	12	1598
S_1	0.5108	13	1043
S_4	0.5259	14	1655
S_{10}	1.5167	15	172

are ordered according to their strength weight. For illustrative purposes, in Energy Services (Fig. 5.9), all parameters have been set equal or close to zero with different probability in the bootstrap samples. Particularly, the stronger weights (over 0.25), most of the times are estimated in the bootstrap networks.

Specifically, the edge V_1 (Competitive salary package) – V_7 (Interesting job) is estimated most of the times as non-zero, since the probability to be zero is only 0.02. On the contrary, the edge V_4 (Long-term job security) – V_8 (Quality of Training) seldom included in the network, as the probability to be zero is very high (0.88). Furthermore, the intervals of the $w_{4,\ 8}$ are depicted wider and more transparent. This signifies that only 13 (over 1500) bootstrap networks estimate the weight $w_{4,\ 8}$; nevertheless, the value of $w_{4,\ 8}$ was not consistent. In some networks, it was negative, while in others was positive ($-0.170 \leq w_{4,\ 8} \leq 0.166$). Overall, the intervals of all weights are wide denoting the questionable accuracy of this network.

On the other hand, Fig. 5.10 presents the quantile intervals of the bootstrapping method for Informatics sector (S_7). The first 20 weights (from V_7–V_8 until V_2–V_3) are estimated in all bootstrap networks since the boxes indicating the percentage of zero is null (0.00). As the weights tend to be lower (close to zero), the percentage to be zero is getting higher. Moreover, the bootstrap weights (black dots) are very close to the true weights (red dots). Finally, 11 edges (from V_6–V_{10} until V_3–V_6) are estimated as absent in the actual network. These weights have been not appeared in the bootstrapped models most of the times (over than 67%). Hence, the negative

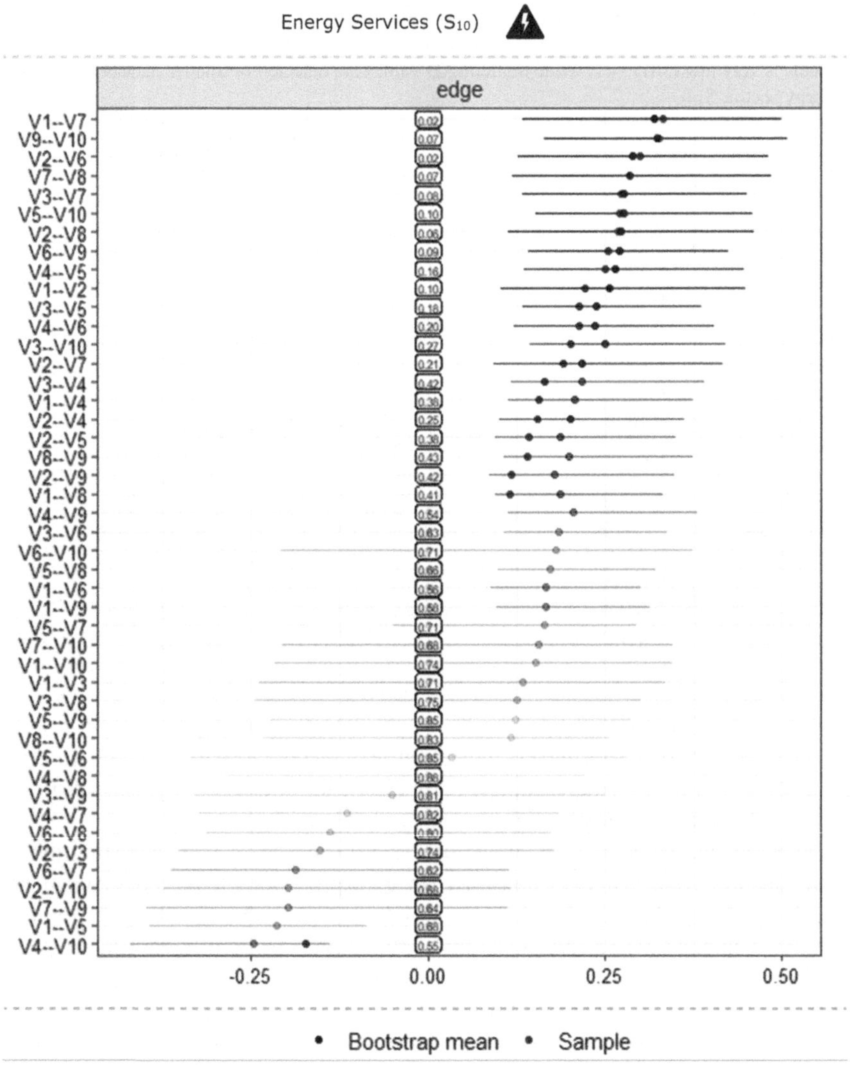

Fig. 5.9 Bootstrapped quantile intervals of each pair of weights in Energy Services (S_{10}). The values in the box indicate the percentage of bootstrapping to be zero. The transparency of the intervals signifies how often an edge is estimated in the bootstrap networks. For instance, the edge V_4–V_8 was estimated as zero in 1320 out of 1500 networks (88%)

weight between V_1 (Competitive salary package) and V_{10} (Dealing with the environment and society) has been estimated as negative in all bootstrap networks. This can be justified a) from the negative quantile intervals ($-0.1514 \leq w_{1,10} \leq -0.0638$) and b) from the zero-percentage box. Overall, it may say that the

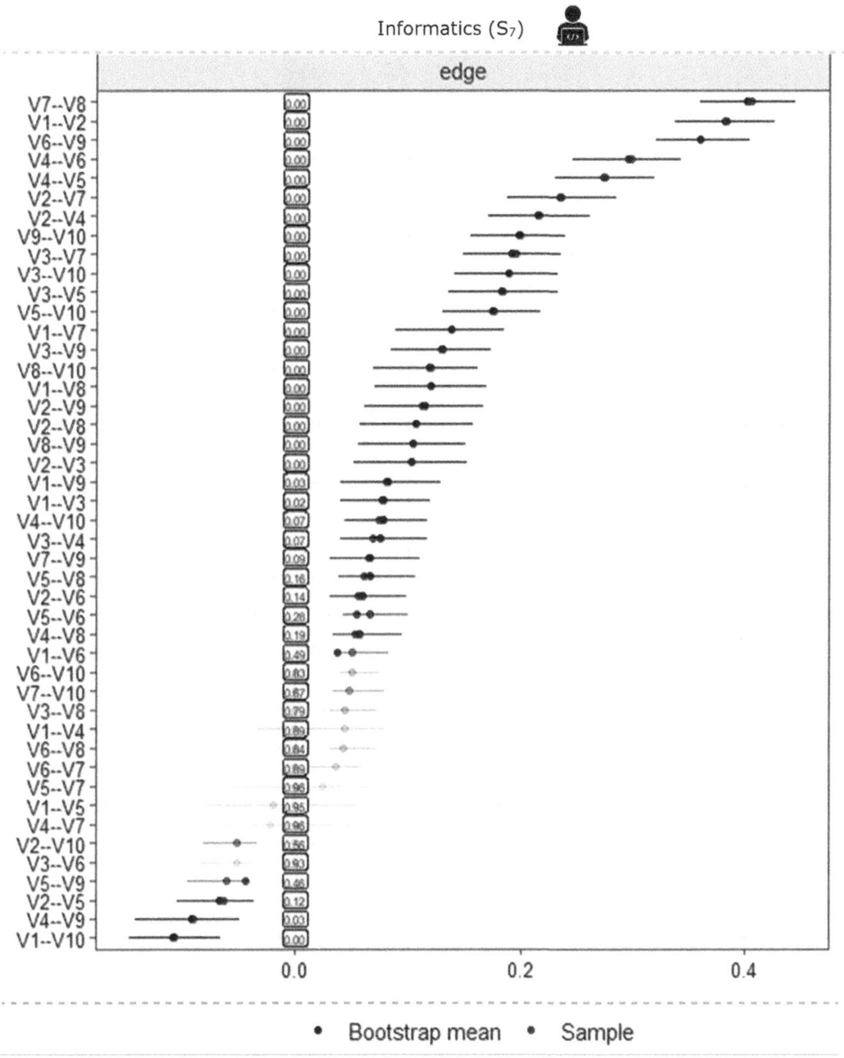

Fig. 5.10 Bootstrapped quantile intervals of each pair of weights in Informatics sector. The bootstrap means and the sample points are very close reflecting an adequate accuracy for the network

Informatics job-satisfaction network is determined as an accurate network since the bootstrap dots and the percentage boxes agree with the true network.

To sum up, the results of this study have been able to demonstrate the accuracy of the networks in all sectors. However, in sectors S_{10}, S_4 and S_1, the accuracy is controversial. Since the sample size plays a significant role in the network accuracy, it is equally important to investigate the stability of centrality indices of these job-satisfaction networks to confirm the robustness of the derived models. Section 5.7 performs the stability of the job-satisfaction networks.

5.7 Network Stability of Job-Satisfaction Networks

After the estimation and analysis of job-satisfaction network models of 15 sectors, it is important to investigate how stable the centrality indices are. To tackle this issue, case-dropping bootstrap is applied to test the overall stability by eliminating gradually observations from the initial dataset. The CS-coefficient concludes the results of the case-dropping bootstrap, which measures the proportion of observations that can be removed from the dataset to retain with 95% probability a correlation of equal or higher than 0.7. This coefficient should be preferable to be higher than 0.5 but not lower than 0.25.

Therefore, the bootnet R package (Epskamp & Fried, 2019) was selected to test the stability of the networks. For each sector, the number of bootstraps is set to 1500. Accordingly, for 15 sectors, 15,000 networks were estimated using the cores (two cores) of the processor (intel(R) Core (TM) i5-4210U CPU@1.70GHz, RAM 8.00GB) to achieve parallel computing. Besides, the estimated networks were based on the same arguments (EBIC, threshold and tuning parameter) that were determined to estimate the true network model (Chap. 4).

Figure 5.11 presents the average correlations between centrality indices of the true network with the centrality indices of 1500 subsamples generated from the original dataset of each sector: Energy Services (S_{10}), Human Resources (S_5), Retail (S_{11}) and Non-profit (S_8). In Energy services (Fig. 5.11a), the network model is not interpretable since the CS-coefficients of all centrality indices indicate poor stability (values are lower than 0.25), $CS(cor = 0.7) \leq 0.052$. Next, CS-coefficients of betweenness and closeness of Human Resources (Fig. 5.11b) do not reach the preferable threshold of 0.5 ($CS(cor = 0.7 = 0.361$); while the CS-coefficients of strength considered as stable ($CS(cor = 0.7 = 0.594$). In Retail (Fig. 5.11c), the orders of closeness ($CS(cor = 0.7 = 0.594$) and strength ($CS(cor = 0.7 = 0.672$) have shown to be stable; while the order of betweenness could be interpretable with some concern ($CS(cor = 0.7 = 0.439$). Last, CS-coefficients of all centrality indices in the Non-profit sector (Fig. 5.11d) seem to be stable and consequently, interpretable ($CS(cor = 0.7 = 0.75$).

Table 5.5 reports the CS-coefficient of centrality indices to denote which network models are stable or in other words, which centrality order could interpret better a model. It is observed that network models of S_1 and S_{10} are unstable; however, S_8 and S_{15} present high stability. Moreover, the order of strength could be interpretable with some care since the CS-coefficient is close to the preferable threshold (CS $(cor = 0.7 = 0.439$). In general, CS-coefficients of strength demonstrate stable models; except for S_3 that indicate the order of betweenness and sectors S_4 and S_6 that show higher stability in the order of closeness.

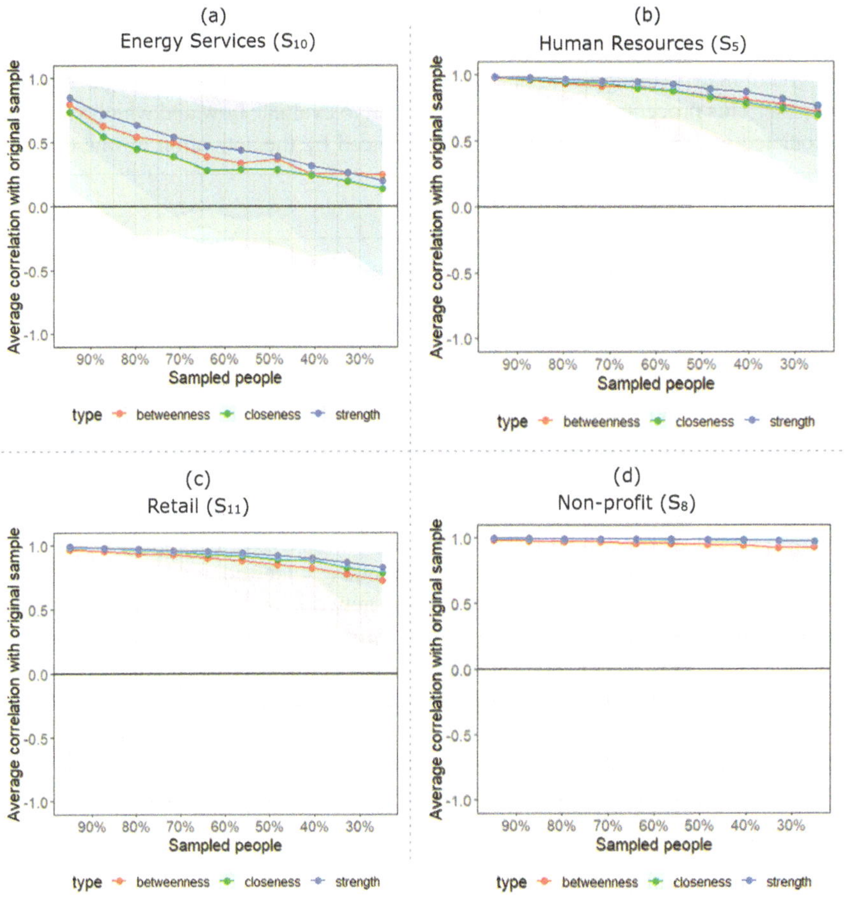

Fig. 5.11 Simulation results using the case-dropping bootstrap in four sectors. Every panel displays the average correlations between centrality indices of the original job-satisfaction networks with the centrality indices obtained from the datasets. Lines represent the average of centrality indices and the colored areas specify the range of the corresponded centrality indices from the 2.5th to the 97.5th quantile

Table 5.5 Maximum drop proportions to retain correlations of 0.7 in at least 95% of the samples of centrality indices, CS(cor = 0.7)

Correlation stability analysis							
	Betweenness	Closeness	Strength		Betweenness	Closeness	Strength
S1	0.128	0.205	0.050	S9	0.517	0.517	0.672
S2	0.283	0.283	0.439	S10	0	0	0.052
S3	0.672	0.594	0.594	S11	0.439	0.439	0.672
S4	0.361	0.595	0.361	S12	0.672	0.594	0.672
S5	0.361	0.361	0.594	S13	0.512	0.672	0.672
S6	0.517	0.594	0.517	S14	0.439	0.594	0.672
S7	0.594	0.439	0.594	S15	0.750	0.750	0.750
S8	0.750	0.750	0.750				

The result of centrality stability does not only explain the interpretability of the model, but it also could be a significant parameter to reorder the variables in the weighted matrix to apply scenario analysis as Fuzzy Cognitive Maps (FCMs) propose. This procedure could be accomplished to evaluate how and which variables associated with job satisfaction can be influenced by the activation of one or more variables in the network (Chap. 6).

5.8 Conclusions

In this chapter, the basic notions of network analysis were presented. Moreover, it was shown how centrality indices were applied to identify the most significant nodes in the network using ordinal datasets. Thus, it was presented how clustering coefficients were applied to identify redundant nodes, in which their removal would not be had a detectable impact on the network. However, the application of network analysis in the business field focused on the job-satisfaction problem which it is in its very infancy. Drawing a parallel with other scientific fields (e.g., physics, genetics, computer science, sociology) where network analysis is more established, we think that this approach has a strong potential to introduce itself in the business field and contribute to advances in empirical and theoretical knowledge.

Finally, we applied nonparametric bootstraps to assess the accuracy of the business network parameters using the bootnet R package (Epskamp & Fried, 2019). All the bootstrap networks were estimated under the same arguments of the true sample network of each sector. Moreover, the total number of bootstrap sample networks was 22,500 (1500 for each sector). The results indicated that the sample size plays a significant role in accuracy since larger samples tend to be more accurate. Nevertheless, for smaller datasets in which the accuracy of the estimated job-satisfaction networks is questionable. To tackle this problem, the stability of centrality indices was employed to confirm the robustness of the derived models using the case-dropping bootstrap. The results showed that the order of strength centrality was interpretable, in all sectors except S_1 and S_{10} due to the low sample size of the datasets.

References

Barabási, A. L., & Bonabeau, E. (2003). Scale-free networks. *Scientific American, 288*, 60–69. https://doi.org/10.1038/scientificamerican0503-60

Barrat, A., Barthelemy, M., Pastor-Satorras, R., & Vespignani, A. (2004). The architecture of complex weighted networks. *Proceedings of the National Academy of Sciences, 101*(11), 3747–3752. arXiv:cond-mat/0311416.

Boccaletti, S., Latora, V., Moreno, Y., Chavez, M., & Hwang, D. (2006). Complex networks: Structure and dynamics. *Physics Reports, 424*(4–5), 175–308.

Borgatti, S. P., & Everett, M. G. (2006). A graph-theoretic perspective on centrality. *Social Networks, 28*(4), 466–484.

Brandes, U. (2001). A faster algorithm for betweenness centrality. *Journal of Mathematical Sociology, 25*(2), 163–177.

Brandes, U. (2008). On variants of shortest-path betweenness centrality and their generic computation. *Social Networks, 30*(2), 136–145.

Chtelat, D., & Wells, M. T. (2016). Improved second order estimation in the singular multivariate normal model. *Journal of Multivariate Analysis, 147*(2016), 1–19.

Costantini, G., Epskamp, S., Borsboom, D., Perugini, M., Mõttus, R., Waldorp, L. J., & Cramer, A. O. J. (2015). State of the aRt personality research: A tutorial on network analysis of personality data in R. *Journal of Research in Personality, 54*, 13–29.

Costantini, G., & Perugini, M. (2016). The network of conscientiousness. *Journal of Research in Personality, 65*, 68–88. https://doi.org/10.1016/j.jrp.2016.10.003

Cramer, A. O. J., van der Sluis, S., Noordhof, A., Wichers, M., Geschwind, N., & Aggen, S. H. (2012). Dimensions of normal personality as networks in search of equilibrium: You can't like parties if you don't like people. *European Journal of Personality, 26*(4), 414–431. https://doi.org/10.1002/per.1866

Csardi, G., & Nepusz, T. (2006). The igraph software package for complex network research. *InterJournal, Complex Systems, 1695.*

Dijkstra, E. W. (1959). A note on two problems in connexion with graphs. *Numerische Mathematik, 1*, 269–271.

Efron, B. (1979). Bootstrap methods: Another look at the jackknife. *The Annals of Statistics, 7*(1), 1–26.

Epskamp, S., & Fried, E. I. (2017). A tutorial on regularized partial correlation networks. *Psychological Methods*. In press. https://doi.org/10.1037/met0000167

Epskamp, S., & Fried. E. I. (2019). *Bootstrap methods for various network estimation routines*. CRAN. http://cran.r-project.org/package=bootnet

Epskamp, S., Borsboom, D., & Fried, E. I. (2017). Estimating psychological networks and their accuracy: A tutorial paper. *Behaviour Research Methods, 2017*, 1–18.

Epskamp, S., Cramer, A., Waldorp, L., Schmittmann, V. D., & Borsboom, D. (2012). qgraph: Network visualizations of relationships in psychometric data. *Journal of Statistical Software, 48* (4), 1–18. http://www.jstatsoft.org/v48/i04/

Erdös, P., & Rényi, A. (1959). On random graphs I. *Publicationes Mathematicae Debrecen, 6*, 290–297.

Foygel, R., & Drton, M. (2010). Extended Bayesian information criteria for Gaussian graphical models. *Advances in Neural Information Processing Systems, 23*, 2020–2028.

Freeman, L. C. (1978). Centrality in social networks conceptual clarification. *Social Networks, 1*(3), 215–239.

Fruchterman, T. M., & Reingold, E. M. (1991). Graph Drawing by force-directed placement. *Software: Practice and Experience, 21*, 1129–1164.

Hastie, T., Tibshirani, R., & Wainwright, M. (2015). Statistical learning with sparsity: The lasso and generalizations.

Holland, P. W., & Leinhardt, S. (1970). A method for detecting structure in sociometric data. *American Journal of Sociology, 76*, 492–513.

Humphries, M. D., & Gurney, K. (2008). Network 'small-world-ness': A quantitative method for determining canonical network equivalence. *PLoS One, 3*(4), e0002051. https://doi.org/10.1371/journal.pone.0002051

Kolaczyk, E. D. (2009a). *Statistical analysis of network data* (pp. 5, 37, 126, 158, 205, 206, 207). Springer.

Kolaczyk, E. D. (2009b). Descriptive analysis of network graph characteristics. In *Statistical analysis of network data*. Springer Series in Statistics. Springer.

Luce, R. D., & Perry, A. D. (1949). A method of matrix analysis of group structure. *Psychometrika, 14*(1), 95–116.

Milgram, S. (1967). The small world problem. *Psychology Today, 2*(1), 60–67.
Murphy, K. P. (2012). *Machine learning: A probabilistic perspective.* MIT Press.
Newman, M. E. J. (2001). Scientific collaboration networks. II. Shortest paths, weighted networks, and centrality. *Physical Review E, 64*, 016132.
Newman, M. E. J. (2004). Analysis of weighted networks. *Physical Review E, 70*(5), 056131. 37, 206, 208.
Newman, M. E. J. (2010). *Networks: An introduction* (pp. 3, 5, 6, 34, 77, 195, 205, 208). Oxford University Press.
Nisbet, R., Miner, G., & Yale, K. (2018). Numerical prediction. *Handbook of Statistical Analysis and Data Mining Applications*, 187–213.
Sabidussi, G. (1966). The centrality index of a graph. *Psychometrika, 31*(4), 581–603.
Shapiro, S. S., & Wilk, M. B. (1965). An analysis of variance test for normality (complete samples). *Biometrika, 52*(3–4), 591–611.
Strogatz, S. H. (2001). Exploring complex networks. *Nature, 410*, 268–276.
Wasserman, S., & Faust, K. (1994). *Social network analysis: Methods and applications.* Cambridge University Press.
Watts, D., & Strogatz, S. (1998). Collective dynamics of small-world networks. *Nature, 393*(6684), 440–442.

Chapter 6
The Proposed Data-Driven glassoFCM Method

Fuzzy cognitive maps (FCMs) have gained popularity within the scientific community due to their capabilities in modeling and decision making for complex problems. However, learning FCM models automatically from data without any expert knowledge and/or historical data remains a considerable challenge. Therefore, we propose a novel algorithm, the glassoFCM (graphical least absolute shrinkage and selection operator and Fuzzy Cognitive Map) to bridge that gap in the literature. Specifically, the glassoFCM is a combination of glasso technique with EBIC (Extended Bayesian Information Criterion) regularization and the FCM methodology. To recap, the glasso is a technique originated from machine learning to model data structures or in simple word to learn the undirected structure of a Gaussian Graphical Model and the FCM method is used to simulate different decision-making scenarios.

The main objective of this chapter is the presentation of a learning method, the glassoFCM which can learn the weights between concepts from the available data without any intervention from experts and literature. In other words, the challenging task in this study is the investigation of the suitable transformation of the weight matrix from a symmetric matrix to be asymmetric to determine the directions of the edges among the concepts and construct the glassoFCM model. The variables are ranked according to four cases: i) strength-centrality ii) closeness-centrality, iii) betweenness-centrality, iv) the average values, v) random order and vi) the inverse strength-centrality. After the simulation process, statistical comparisons are applied to examine if there are significant differences in the value of the output concept (V_{11}) when the input concepts are rearranged according to six cases. In total, 90 stable models are estimated revealing the job-satisfaction network structure of 13 working sectors and 1170 scenarios are simulated to indicate not only the deviation of the output values but also to ascertain the most important factors that influence the output concept ('willingness to work') of the FCM. In order to examine and estimate the FCM inference, we applied the 'fcm' R package.

Z. Dikopoulou, *Modeling and Simulating Complex Business Perceptions*, Fuzzy Management Methods, https://doi.org/10.1007/978-3-030-81496-0_6

6.1 Introduction

Generally, FCM models are categorized into three types: i) manual, ii) semi-automated and automated models. Manual FCMs are produced by experts, semi-automated FCMs are constructed by a relatively expert intervention and automated FCMs are estimated from historical data. Expert-based models require relatively less data and lead to simpler and sparser interpretable models (Stach et al., 2012; Amirkhani et al., 2017; Dikopoulou et al., 2020). On the other hand, data-driven models require a sufficient amount of training historical data; however, they are more expressive since they can uncover unexpected patterns of the data comparing to expert-based models (Stach et al., 2012; Papageorgiou & Salmeron, 2013; Papageorgiou, 2014). Mainly, the FCM learning methods which construct and train an FCM fall into three categories: i) the adaptive-based (usually adapted from the Hebbian law), ii) the population-based and iii) the hybrid learning algorithms (Stach et al., 2012; Papageorgiou & Salmeron, 2013; Amirkhani et al., 2017). The adaptive-based algorithm is a semi-automated method since it updates the weights of the connection matrix obtained from experts. Population-based (evolutionary) learning algorithm is an automated method which uses solely historical data to estimate the weights between concepts. The main goal of this algorithm is the estimation of an appropriate model that miming the input data. Finally, the hybrid approach combines the adaptive and population-based learning methods in which the knowledge of an expert and historical data is simultaneously used.

Therefore, most of the studies have been focused on semi-automated and automated methods to learn FCM based on historical data. However, few approaches have been applied to define the relations and the weights among concepts (nodes) from cross-sectional data (not historical data). In general, a cross-sectional study is defined as an observational research in which multiple variables can be examined at a given point in time (Knowland et al., 2015).

As mention in previous chapters, concerning datasets, two well-known methods were used to construct FCMs, SEM (Bollen, 1989) and distance-based method (Schneider et al., 1998). In the introduction of Chap. 4, it was mentioned that two important conditions were necessary to create an SEM, data sufficiency and high clustering. However, in this study, the majority of the methods determine that the cumulative variance of the variables of all datasets could be satisfactorily explained with only one cluster (Sect. 4.5). On the other hand, as described in Chap. 4, the distance-based method defines associations among concepts but expert intervention is important to indicate the direction of the edges. Nevertheless, we prove that the distance-based approach is inefficient in large datasets since it provides complete graphs with high weights between variables. Moreover, we have investigated that it is impossible to make decisions and policies to predict the most important parameters enhancing the companies' attractiveness using the distance-based method (Table C.2). This is due to the fact that there are no deviations on the calculated values of the concepts for the conducted FCM scenarios (Dikopoulou et al., 2017; Table C.2).

Nevertheless, an automated FCM model estimation from (cross-sectional) data assessing the direct relationships between concepts is a major challenge. In order to bridge the gap, we propose an automated data-driven method, the *glassoFCM* (Dikopoulou et al., 2017) to model an FCM (concerning the job satisfaction) from categorical ordinal observations without any expert intervention and/or usage of historical data. The glassoFCM is a combination of two methods, the *graphical least absolute shrinkage and selection operator* (glasso) with *EBIC* (*Extended Bayesian Information Criterion*) regularization for data modeling (Chap. 4) and the FCM simulation for decision making (Chap. 3). Consequently, the weighted matrix is derived from glasso method which it is regularized with the EBIC technique and it simulated using the inference rules and threshold functions from FCM methodology.

The obstacle in this study is the identification of the directions in the graphs since the glasso method returns symmetric weighted matrices (e.g. Table 4.6) and expert knowledge to indicate the direction of the edges is absent. One solution is to use the upper triangular part of the weighted matrix to run the FCM scenarios by changing the order of the variables. This procedure indicates that the first variable in the matrix stimulates (initiates) the remaining concepts in the map (since the in-degree index is zero) while the last variable of the part above the diagonal of the weighted matrix will be always the receiver (out-degree equals to zero). Consequently, hierarchical structures will be produced. Therefore, it is crucial to choose the appropriate node (transmitter) to be placed in the first position of the upper-triangular weighted matrix since it can influence the remaining nodes of the glassoFCM model. The variables are ranked according to six cases: i) strength-centrality ii) closeness-centrality, iii) betweenness-centrality which these indices derived from the output matrix from glasso method, iv) the average which derived from the dataset, v) random order and vi) the inverse strength-centrality. In other words, only the results of the decision output concept ("Willingness to work") are inspected. This implies that the focal concept is always placed in the last (p^{th}) position of the upper-triangular weight matrix; where p represents the total number of the concepts. The remaining ten variables represent the input concepts associated with job-satisfaction features and they are rearranged according to the ranking cases. Several experiments are conducted and the results are validated to justify if the observed rearrangements play a determinant role in the output values. Moreover, we determine which of the variables are the most significant features that increase the target variable ("Willingness of the employees to work").

To sum up, Sect. 6.2 introduces the glassoFCM, a combined method of glasso with EBIC regularization to model data structures and FCM to simulate the derived model. In Sect. 6.3, the stability of new job-satisfaction models using eleven nodes/concepts is tested. The results of glassoFCM method are applied to the job-satisfaction problem and they are presented in Sect. 6.4. Moreover, Sect. 6.4.1 shows the significant variables that influence the target variable, Sect. 6.4.2 presents the statistical comparisons of the FCM simulation and finally, Sect. 6.5 outlines the conclusions.

6.2 The Graphical Lasso FCM Algorithm (glassoFCM)

The glassoFCM methodology (Dikopoulou et al., 2017) is a combination of the regularized glasso algorithm (Chap. 4) with the FCM method (Chap. 3) to model graphs from large datasets and simulate different decision-making scenarios (Fig. 6.1).

First, glassoFCM algorithm is used to model a network structure directly from a dataset. Particularly, the glasso method with the EBIC regularization is applied as described in Chap. 4 (Algorithm 4.1). Since the tuning parameter λ (lambda) is equal to zero this indicates that 100 different models are created from fully-connected models ($\lambda = 0$) to empty models ($\lambda = \infty$). Then, the model that minimizes the EBIC is selected. Therefore, a symmetric weighted matrix is produced. Then, the direction of the edges between nodes is determined by obtaining the upper-part of the diagonal of the weighted matrix.

If expert knowledge is available, then edges are eliminated from the weighted matrix to create the directed weighted matrix (W). For every decision an expert makes, a scenario-policy is set. Consequently, an initial vector (V^0) is created by activating one or more concepts depending on the hypothesis (scenario). According to the problem and/or the expert experience, one of the six inference rules is selected to update the initial vector. To reduce the unbounded input of V to a strict range that threshold controls, the threshold function is chosen. The vector is updated and the time-step (κ) is increased by one. As occurred in the original FCM inference, the simulation stops when a) the system reaches the equilibrium point, b) reaches a limited number of time-steps or c) a chaotic behavior occurs. Below, the glassoFCM algorithm is described.

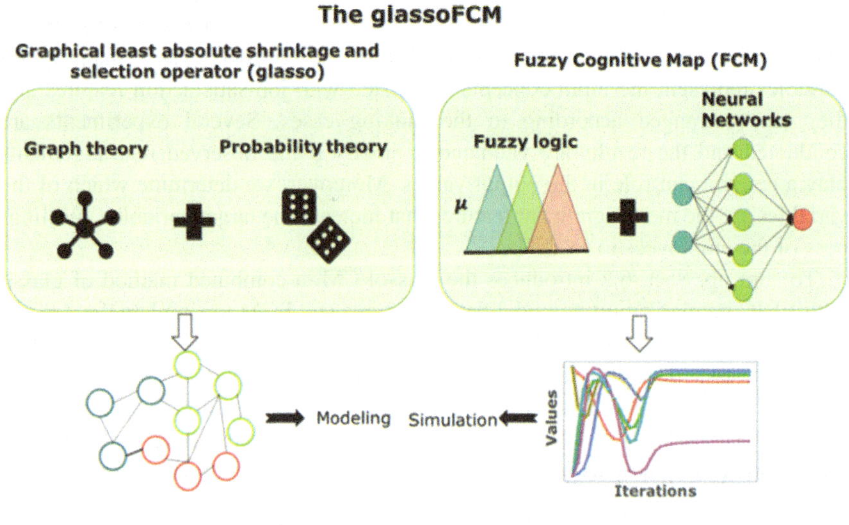

Fig. 6.1 The data-driven glassoFCM algorithm which models and simulated cross-sectional datasets without any expert knowledge or historical data

Algorithm 6.1: The glassoFCM with EBIC regularization using the inverse covariance matrix

Input Dataset (n $\times$ p, n: observations, p: variables), $\lambda_{\aleph}$ (vector of 100 values), κ_{max} (maximum number of steps), C^0(initial vector, 1 $\times$ p), ε (residual error)

Output A converged vector $V^{\kappa} = \left[V_1^{\kappa}, V_2^{\kappa}, \ldots, V_p^{\kappa}\right])$

Step 1: **For** each $\lambda_{\aleph}$
Step 2: Initialize $\mathbf{W} = \mathbf{S} + \lambda\mathbf{I}$. The diagonal of W remains unchanged in what follows
Step 3: Partition the matrix W into two parts. Part 1 includes all columns except p^{th} row and p^{th} column and part 2 consists of the pth row and column. Perform the following steps till convergence:
Step 4: Rearrange the rows/columns so that the target column (part 2) is last
Step 5: Solve the lasso problem (4.17), using as initial value the solution from the previous round for this column.
Step 6: Update the row/column (off-diagonal) of the covariance using $\widehat{w}_{12}$ (4.15).
Step 7: Save $\widehat{\beta}$ for this column in the matrix B.
Step 8: Finally, for every row/column, calculate the diagonal entries $\widehat{\theta}_{jj}$ using (4.18), and transform the B matrix to Θ.
Step 9: Calculate the EBIC in Equation 4.19.
End For
Step 10: Choose the graph G that minimizes EBIC (the symmetric weight W matrix is estimated)
Step 11: Get the upper-triangular matrix of W
Step 12: **While** $\kappa \leq \kappa_{max}$ AND $\left(V_i^{\kappa} \neq V_i^{(\kappa+1)} \text{ or } V_i^{\kappa} - V_i^{(\kappa+1)} > \varepsilon\right)$ **do**:
Step 13: Apply the inference rule (3.1–3.3) to update the initial Vector V^0.
Step 14: Use the threshold function (3.4–3.7) to reduce the unbounded input of V to a strict range that threshold controls.
Step 15: Save the updated $V^{(\kappa+1)}$ (consider that this vector will be the initial vector κ for the next iteration).
Step 16: Increase step κ ($\kappa \leftarrow \kappa + 1$)
End while
Step 17: Return the converged vector V^{κ}
END

6.3 Evaluating the Network Stability of the New Job-Satisfaction Models

In the job-satisfaction problem, the "Willingness of the employee to work in the company" is added and it is considered the output variable of the model. Before we apply the glassoFCM algorithm, it is necessary to examine the stability of the new models.

In total, 15 new models are created, each for every working sector. From the case-dropping bootstrap method, the Automotive (S_1) and the Energy services (S_{10}) sectors are excluded from the study since they present low stability (*CS(coeff)* < 0.5; Table 6.1). Nevertheless, the rest working sectors show an increment of the CS-coefficient comparing to Table 5.5 indicating that the glassoFCM can be used in high-dimensional datasets.

Table 6.1 Maximum drop proportions to retain correlations of 0.7 in at least 95% of the samples of centrality indices, CS(cor = 0.7) in job-satisfaction networks of 11 variables. Automotive (S_1) and Energy services (S_{10}) are removed from this study since they are unstable

Correlation stability analysis of 11 variables							
	Betweenness	Closeness	Strength		Betweenness	Closeness	Strength
S1	0	0.284	0.284	S9	0.672	0.672	0.750
S2	0.128	0.595	0.672	S10	0.052	0	0.052
S3	0.534	0.750	0.75	S11	0.361	0.672	0.672
S4	0.310	0.675	0.675	S12	0.750	0.750	0.750
S5	0.439	0.594	0.672	S13	0.594	0.750	0.750
S6	0.672	0.750	0.750	S14	0.672	0.750	0.750
S7	0.672	0.594	0.750	S15	0.750	0.750	0.750
S8	0.750	0.750	0.750				

6.4 The Results of the glassoFCM in Job-Satisfaction Problem

As mentioned above, the V_{11} variable is considered the output concept of the glassoFCM model. For this reason, it is placed in the last position of the dataset. Consequently, the number of columns of each dataset is increased by one, while the number of rows/observations is the same, for instance, the dimensions of the Finance (S_3) dataset is 11 × 3819. The remaining concepts $V_1 - V_{10}$ are rearranged according to six cases/categories: i) strength, ii) closeness, iii) betweenness, iv) average of variables in the datasets, v) random and vi) the inverse of the strength. Therefore, in every sector, 90 scenarios are simulated, 15 for each category. In total, for the 13 sectors, 1170 simulations are performed. After reordering the variables in the dataset, the glasso algorithm is applied to obtain the weighted matrix (11 × 11). Then, the upper triangular part is used to determine the directions of the edges. Next, the simulation of the glassoFCM is applied to investigate the influence of all concepts into V_{11} (target concept) according to the initial activating values of the concepts of the initial vector. To recall, every *what-if* scenario creates an initial vector in which one or more concepts (nodes) are activated. For this research, the procedure of simulation is repeating for six cases.

Figure 6.2, presents the steps of the proposed procedure to utilize the FCM simulation and compare the values of the target variable (V_{11}). In step 1, the initial data is set and the average values of the variables are calculated. The variables are ranked according to the mean value and this order is saved in the sixth step. Next, the glasso method with EBIC regularization technique is applied to the dataset to identify the best model that fits into the data (step 2).

The weight matrix derived from the glasso is (step 3). Bootstrap methods are applied for testing the accuracy and the stability of the network (step 4). If the network is inaccurate and unstable, then the network is out of the research. Alternatively, if the model is accurate and stable, centrality measures will be defined. The centrality rankings of the z cases are stored (step 5) including the average ranking,

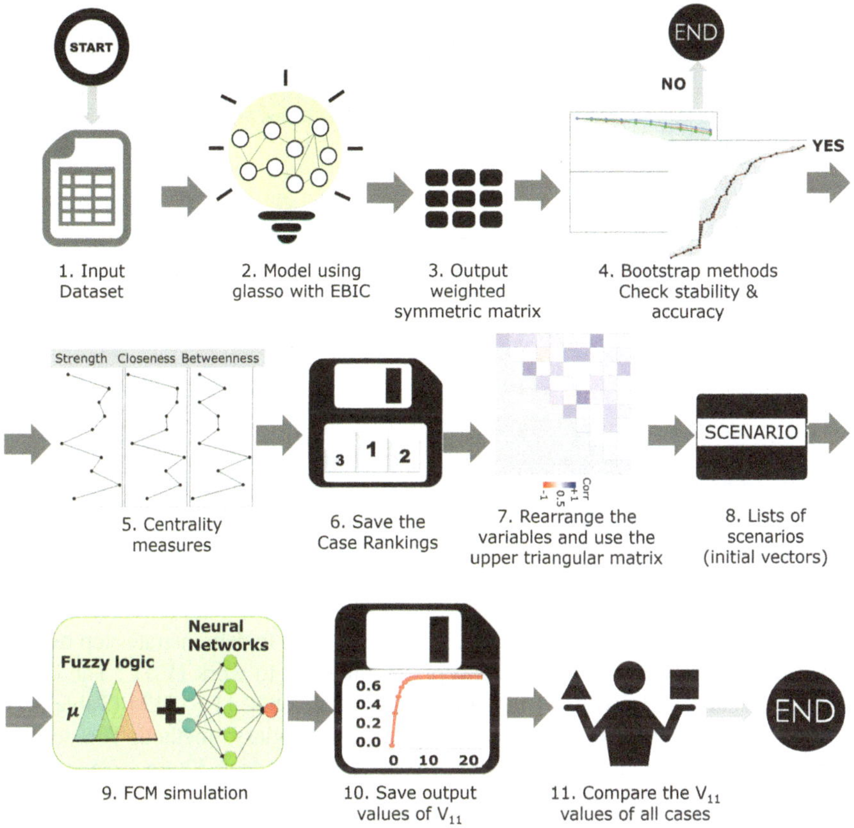

Fig. 6.2 The procedure to obtain FCM simulations from graphical lasso models (steps 1–10). Moreover, the output values of V_{11} are compared to validate if they are statistically significantly different in the observed cases (step 11)

the random reordering and the inverse strength centrality. In this study, the number of examined cases are $z = 6$. It is notable to mention that reordering the variables in the dataset, the weights among the nodes derived from glasso method are not changing. Therefore, the glasso with EBIC regularization technique will be applied only once. Thus, the upper triangular part is selected from the produced weight matrix (step 7). Afterwards, in step 8, the initial vectors are determined ($N\_scenarios = 13$). Then, the 'fcm.infer' function from 'fcm' package (Dikopoulou & Papageorgiou, 2017; Dikopoulou et al., 2018) is applied for the simulation procedure (step 9). The output value of the target concept is saved in a matrix of, $N\_scenarios \times z$ (step 10). When the scenarios for one possible case are completed, then weights in the upper triangular matrix are reordered according to the next case as step 6 indicates. Moreover, when the output values of V_{11} are performed for all cases then the values are compared in the last step (step 11).

Table 6.2 A network analysis of three hierarchical directed FCMs

ATTRIBUTES	FINANCE (S_3)	NON-PROFIT (S_8)	BUSINESS (S_{12})
Observations	3819	20378	3134
Receiver	1	1	1
Transmitter	1	1	1
Ordinal	9	9	9
Connections	42	49	34
Positive Connections	38	42	32
Negative Connections	4	7	2
Connections to V11	7	9	3
Hierarchy index (h)	0.8148	0.7884	0.5901

For the FCM simulation, the Business dataset is used to demonstrate step by step the estimated connections of the concept "Willingness to work (V_{11})". The cross-sectional dataset consists of 11 variables and 3134 records, 11 × 3134 (Table 6.2). Then, the job-satisfaction model is estimated with the glasso method. The derived network structure obtains 34 connections (out of 55) in which 32 are assigned with positive weights and 2 with negative weights (Fig. 6.3b). Specifically, the highest positive weights are observed between Competitive package and Career Opportunities ($V_1 - V_2$, $w_{12} = .389$) and between Interesting jobs and Quality of training ($V_7 - V_8$, $w_{7,\,8} = .389$).

On the contrary, the negative weights are emerged among Long term job security and Strong management ($V_4 - V_9$, $w_{4,\,9} = -.076$) and among Good balance between private life and work and Strong management ($V_5 - V_9$, $w_{5,\,9} = -.065$) indicating that the increase of V_9 causes a very low decrement of V_4 and V_5 and vice versa. Then, the case-dropping bootstrap method is applied to test the stability of the model (Fig. 6.3c). After 1500 bootstrap sample networks, the results show that the business sector is stable; particularly, in the strength-centrality index, $CS_{S_{1,2}}^{Str}(cor = 0.75)$ and $CS_{S_{1,2}}^{Bet} = CS_{S_{1,2}}^{Clos}(cor = 0.672)$. Next, the centrality indices are computed (Fig. 6.3d) to reorder the weights in the weighted matrix.

The rankings of all cases are included in a matrix where the output concept (V_{11}) is placed in the last position. Figure 6.3e presents the ranked cases of strength, closeness, betweenness, mean, random ad inverse strength that will be used to arrange the weights. Afterwards, the weights of the upper triangular matrix are reordering according to the first case of ranking as Fig. 6.3f depicts. Before the FCM simulation process, it is necessary to determine the initial vector(s) V that includes the activated values of all or the selected concepts. The table in Fig. 6.3g

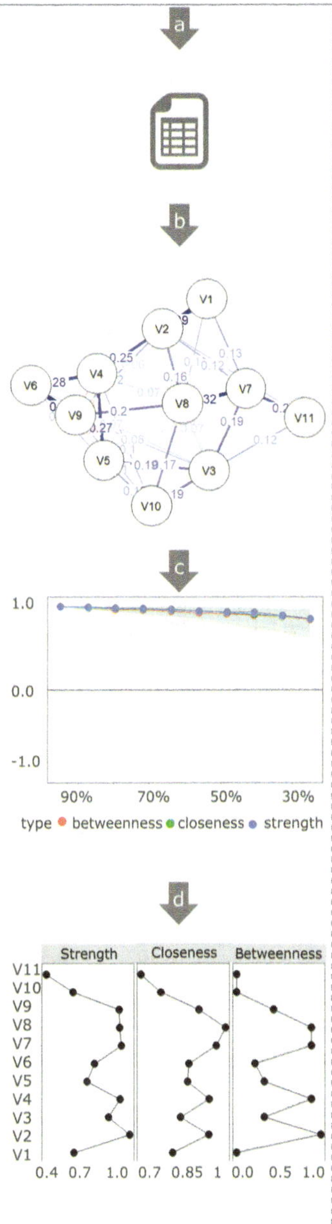

(e)

Reordering cases in Business sector

RANK	Strength	Closeness	Between-ness	Mean	Random	Reverse_strength
1st	V2	V8	V2	V6	V1	V10
2nd	V7	V7	V4	V9	V2	V1
3rd	V4	V4	V7	V4	V3	V5
4th	V8	V2	V8	V2	V4	V6
5th	V9	V9	V9	V1	V5	V3
6th	V3	V6	V3	V7	V6	V9
7th	V6	V5	V5	V8	V7	V8
8th	V5	V3	V6	V3	V8	V4
9th	V1	V1	V1	V5	V9	V7
10th	V10	V10	V10	V10	V10	V2
11th	V11	V11	V11	V11	V11	V11

(f)

V2	V7	V4	V8	V9	V3	V6	V5	V1	V10	V11
0	0.121	0.250	0.162	0.118	0.051	0	0	0.389	0	0.082
	0	0	0.324	0	0.186	0.042	0.054	0.129	0	0.245
		0	0.070	-0.076	0.036	0.285	0.274	0	0.097	0
			0	0.199	0.065	0	0	0.101	0.166	0
				0	0.084	0.377	-0.065	0.064	0.099	0
					0	0.066	0.190	0	0.190	0.117
						0	0.091	0	**0**	0
							0	0	0	0
								0	0	0
									0	0
										0

(g)

Scenarios

	Finance	Non-profit	Business
1st	V2 =1	V9 =1	V2 =1
2nd	V4 =1	V2 =1	V7 =1
3rd	V9 =1	V4 =1	V4 =1
4th	v5 =1	V7 =1	V8 =1
5th	V9=V7=1	V8 =1	V9 =1
6th	V2=v9=1	V2=v9=1	V2=V7 =1
7th	V2=V7=1	V4=v9=1	V2=V4 =1
8th	V2=v5=1	V7=V9=1	V2=V8 =1
9th	V2=v4=1	V4=v7=1	V7=V4 =1
10th	V2=v9=V5=1	V2=v7=1	V7=V8=1
11th	V2=V9=V4=1	V2=V9=V4=1	V2=V4=V7 =1
12th	V2=V4=V7 =1	V2=V4=V7=1	V2=V7=V8 =1
13th	V2=v9=V7=V4=1	V4=V7=V9=1	V7=V4=V8 =1
14th	V1=V2=V3=V4= V6=V7=V10=1	V1=V2=V3=V4= V5=V7=V8=V9= V10=1	V2=V3=V7=1
15th	ALL V=1	ALL V=1	ALL V=1

Fig. 6.3 The pre-processing of the Business sector to identify the initial vectors and the initial matrix for the FCM simulation

illustrates 15 scenarios of three sectors (Finance, Non-profit and Business). Each row of this table represents a scenario of each sector. The last column determines the scenarios of the business sector. The interpretation of the scenarios is simple; for instance, the first scenario ascertains that V_2 is the triggering concept while the remaining concepts in the map are inactive (the initial values are set equal to zero).

The aim of this procedure is the comparison of the output values (derived from the FCM method) of the proposed cases. Consequently, if the values are not statistically significantly different, then it is inferred that rearranging the concepts in the hierarchal FCM does not play a considerable role in the simulation.

However, it is important to emphasize that in the simulation process, only the results of one output variable (V_{11}) will be observed which it is placed in the last position of the weight matrix.

6.4.1 The Significant Variables in the FCM Structures

In order to indicate which variables can be activated to achieve the highest output values in hierarchy structures, a different number of scenarios were conducted (Table 6.3). As a result, the highest value of V_{11} is observed when its neighbor nodes are activated. For instance, in the Business sector, when the neighbors of V_{11}

Table 6.3 The simulation of the V_{11} values for the six observed cases in the Business sector

		Output of V11					
	Scenarios	**Strength**	**Closeness**	**Between-ness**	**Mean**	**Random**	**Reverse_ strength**
1)	**V2**	0.561	0.544	0.561	0.566	0.559	0.539
2)	**V7**	0.628	0.632	0.628	0.628	0.618	0.622
3)	**V4**	0.502	0.516	0.502	0.518	0.502	0.509
4)	**V8**	0.504	0.554	0.504	0.503	0.499	0.545
5)	**V9**	0.504	0.505	0.504	0.512	0.499	0.512
6)	**V2-V7**	0.666	0.665	0.666	0.665	0.656	0.653
7)	**V2-V4**	0.562	0.549	0.562	0.568	0.562	0.539
8)	**V2-V8**	0.564	0.588	0.564	0.568	0.559	0.576
9)	**V7-V4**	0.630	0.645	0.630	0.639	0.618	0.630
10)	**V7-V8**	0.631	0.643	0.631	0.630	0.618	0.628
11)	**V2-V7-V4**	0.667	0.669	0.667	0.667	0.656	0.653
12)	**V2-V7-V8**	0.668	0.669	0.668	0.667	0.656	0.653
13)	**V7-V4-V8**	0.632	0.653	0.632	0.641	0.618	0.634
14)	**V2-V3-V7**	0.699	0.699	0.699	0.699	0.699	0.699
15)	**ALL Vi**	0.698	0.698	0.698	0.698	0.698	0.698

Table 6.4 The comparisons of the output values (V_{11}) to determine which of the observed activations achieves the highest output value. The variables are initiated according to a) the neighbors of the output variable and b) the highest mean variables. The validations are accomplished by filtering the initial dataset

	Finance		Non-Profit		Business	
	Neighbours of V11	Variables according to MEAN	Neighbours of V11	Variables according to MEAN	Neighbours of V11	Variables according to MEAN
	V1=V2=V3=V4=V6=V7=V10=1	V1=V2=V4=V6=V7=V8=V9=1	V1=V2=V3=V4=V5=V7=V8=V9=V10=1	V1=V2=V3=V4=V5=V6=V7=V8=V9=1	V2=V3=V7=1	V4=V6=V9=1
Strength	0.7445	0.6969	0.7311	0.7181	0.6988	0.5064
Closeness	0.7444	0.6978	0.7311	0.7181	0.6988	0.5228
Betweenness	0.7446	0.6992	0.7311	0.7181	0.6988	0.5061
Mean	0.7444	0.7012	0.7315	0.7181	0.6988	0.5395
Random	0.7445	0.7057	0.7315	0.7180	0.6988	0.5061
Reverse_Strength	0.7444	0.6781	0.7314	0.7037	0.6988	0.5322
Filtered_Mean	5	5	5	4.7143	4.0338	3.659

are activated ($V_2 = V_3 = V_7 = 1$) then the maximum influence is attained (14th scenario of Table 6.3) which is similar to the last scenario (15th), in which all concepts are initiated. Specifically, the difference between two scenarios (14th and 15th) in the Business sector is negligible (.00039 ± .00000); as well as in the rest sectors, in Finance (.00038 ± .00007) and Non-Profit (.00019 ± .00021).

Nevertheless, many people believe that the significant variables that affect significantly the willingness of the employees to work are the ones that obtain higher arithmetic mean values. To respond to this debate, first, the number of neighbors of the output value is counted. For example, in the business sector, the neighbors of V_{11} are three (Fig. 6.3b, f); therefore, the top three ranked variables according to mean values will be activated (Fig. 6.3e) which are: Financially sound (V_6), Strong management (V_9) and Long-term job security (V_4). Next, the variables are rearranged according to the observed cases that were explained in Sect. 6.4 and the fcm.infer function is applied. Each case returns an output value (Table 6.4), $V_{11}^{14} = 0.5064$ (strength), $V_{11}^{14} = 0.5228$ (closeness), $V_{11}^{14} = 0.5061$ (betweenness), $V_{11}^{14} = 0.5395$ (mean), $V_{11}^{14} = 0.5061$ (random) and $V_{11}^{15} = 0.5322$ (inverse_strength). Each of these results makes an important contribution to our understanding of the output values are lower ($V_{11}^{15} < 0.6988$; Table 6.4) when variables are activated according to their mean value compared to the outputs when the neighborhood of V_{11} is triggered). Accordingly, activating the top-7 and top-9 mean variables in Finance and in Non-Profit, respectively, the output values are lower comparing to the output values when its neighbors are initiated (Table 6.4).

To validate the results of the FCM simulation, the variables of the initial cross-sectional dataset are filtered with the highest values of the Likert scale ('5'). This procedure will be repeated twice; i) for the neighbors of V_{11} and ii) for the variables with the highest mean value. Consequently, in the Business sector, for the first occasion, V_2, V_3 and V_7 and for the second occasion, highest mean variables V_4, V_6 and V_9 will be filtered equally to five. From the last row of Table 6.4, it is deduced that the filtered-mean value of V_{11} in the first occasion is higher than the second one (with the only exception in the Finance sector that both occasions return five).

To conclude, in hierarchical FCMs where the output is known, if the neighbors of the output concept are activated, then the output concept will obtain the highest value of the FCM simulation process.

6.4.2 Statistical Comparisons of the FCM Simulations

Previously, through the modelling and the simulation process, we identified the most significant job-satisfaction variables. This indicates that activating the neighborhood of the output (the concepts that are connected directly with the output) then the willingness of the employees to work will be improved. Specifically, the output will obtain the highest value comparing to other scenarios. In this subchapter, we will investigate if the rearrangements of the variables of the observed cases play a significant role in the output value.

Before the comparisons, it is crucial to measure the normality of the data to use the appropriate comparison test. For small sample sizes (<50 samples), the Shapiro-Wilk test (Shapiro & Wilk, 1965) is a more relevant test to assess normality. If the p-value is less than or equal to 0.05 then the test rejects the hypothesis of normality

Table 6.5 Normality Test of three sectors. As the Sig. value (p-value) under the Shapiro-Wilk column is greater than 0.05, it is concluded that the observations of the correspondent case are normally distributed

Tests of Normality - Shapiro-Wilk

	Finance (S_3)			Non-Profit (S_8)			Business (S_{12})		
	Statistic	df	Sig.	Statistic	df	Sig.	Statistic	df	Sig.
Strength	0.943	15	0.428	0.940	15	0.379	0.895	15	0.081
Closeness	0.955	15	0.602	0.913	15	0.150	0.900	15	0.096
Betweenness	0.933	15	0.301	0.906	15	0.117	0.895	15	0.080
Average	0.938	15	0.358	0.934	15	0.317	0.908	15	0.127
Random	0.928	15	0.255	0.949	15	0.514	0.914	15	0.155
Reverse_Strength	0.874	15	0.039	0.940	15	0.388	0.911	15	0.143

Table 6.6 The output of the ANOVA analysis of (a) Finance and (b) Non-Profit show that the differences in the cases are not significant. Additionally, a Kruskal-Wallis H test determined that there was a statistically significant difference since $p > 0.05$

(a)

Kruskal-Wallis Test

	Finance (S3)		
	Chi-Square	**df**	**Asymp. Sig.**
Test Statistics	3.639	5	0.602

(b)

ANOVA

	Non-Profit (S8)				
	Sum of Squares	**df**	**Mean Square**	**F**	**Sig.**
Between Groups	0.002	5	0.000	0.075	**0.996**
Within Groups	0.383	84	0.005		
Total	0.384	89			

(c)

ANOVA

	Business (S12)				
	Sum of Squares	**df**	**Mean Square**	**F**	**Sig.**
Between Groups	0.004	5	0.001	0.104	**0.991**
Within Groups	0.593	84	0.007		
Total	0.597	89			

and the non-parametric test will be used. Otherwise, if the p-value is greater than 0.05, then the outputs are normally distributed and a parametric test will be applied. Table 6.5 presents the results from the Shapiro-Wilk test. The output values derived from FCM of the six observed cases, in Non-profit and in Business sectors, the values are normally distributed because of the Sig. value (p-value) of the Shapiro-Wilk test is greater than 0.05. In Finance, the Sig. value in inverse_strength is below than 0.05; therefore, the data significantly deviate from a normal distribution.

The data in Non-profit and in Business sectors are normally distributed, the dependent variable is continuous and the independent variable consists of six categorical individual groups. For that reason, the appropriate method to determine whether there are any statistically significant differences between the means of six unrelated cases is the one-way analysis of variance (ANOVA; Fisher, 1918). Table 6.6b, c include the ANOVA tables. Specifically, the significance values in Non-profit ($F(5, 84) = 0.075, p = .996$) and in Business ($F(5, 84) = 0.104, p = .991$) are above 0.05.

As a consequence, there are no statistically significant differences in the mean values of V_{11} between the different cases for S_8 and S_{12}. Moreover, determining

Table 6.7 Test statistics table. Tukey multiple comparisons test is applied when data are normally distributed (Finance and Non-profit sectors) and the Mann-Whitney test is used when data significantly deviate from normality (Business sector)

		Mann-Whitney Test				Multiple Comparisons - Tukey HSD					
		Finance (S3)				Non-Profit (S8)			Business (S12)		
Group (I)	Group (II)	Mean Rank Group I	Mean Rank Group II	Mann-Whitney U	Asymp. Sig. (2-tailed)	Mean Difference	Std. Error	Sig.	Mean Difference	Std. Error	Sig.
Strength	Closeness	15.83	15.17	107.50	0.836	0.011	0.031	0.999	-0.007	0.025	1.000
	Betweenness	15.90	15.10	106.50	0.803	0.005	0.031	1.000	0.000	0.025	1.000
	Average	16.50	14.50	97.50	0.534	-0.006	0.031	1.000	-0.003	0.025	1.000
	Random	17.13	13.87	88.00	0.309	0.010	0.031	0.999	0.007	0.025	1.000
	Inv_Strength	17.50	13.50	82.50	0.213	0.010	0.031	0.999	0.002	0.025	1.000
Closeness	Betweenness	15.17	15.83	107.50	0.836	-0.007	0.031	1.000	0.007	0.025	1.000
	Average	16.13	14.87	103.00	0.693	-0.018	0.031	0.993	0.004	0.025	1.000
	Random	16.77	14.23	93.50	0.431	-0.001	0.031	1.000	0.014	0.025	0.993
	Inv_Strength	17.73	13.27	79.00	0.164	-0.001	0.031	1.000	0.009	0.025	0.999
Between-ness	Average	16.37	14.63	99.50	0.590	-0.011	0.031	0.999	-0.003	0.025	1.000
	Random	17.17	13.83	87.50	0.300	0.005	0.031	1.000	0.007	0.025	1.000
	Inv_Strength	17.43	13.57	83.50	0.229	0.005	0.031	1.000	0.002	0.025	1.000
Average	Random	16.27	14.73	101.00	0.633	0.017	0.031	0.994	0.010	0.025	0.998
	Inv_Strength	17.33	13.67	85.00	0.254	0.016	0.031	0.995	0.005	0.025	1.000
Random	Inv_Strength	17.03	13.97	89.50	0.340	0.000	0.031	1.000	-0.005	0.025	1.000

which groups differ from each other, Tukey post-hoc test (Tukey, 1949) is applied. The results in Table 6.7 indicate that between cases, there are no statistically significant differences in sectors S_8 and S_{12}. Particularly, the p-value is close to 1 ($p \geq .993$) indicating that all pairs of cases reach the perfect similarity.

Nevertheless, in the Finance sector (S_3) since the data are not normally distributed; therefore, the Kruskal-Wallis H nonparametric test (Kruskal & Wallis, 1952) is applied to identify statistically significant differences between the cases. A Kruskal-Wallis H test (Table 6.6a) shown that there was no statistically significant difference in V_{11} values between the observed cases, $\chi^2(5) = 3.693$, $p = 0.602$. Moreover, to identify the differences among cases, the Mann-Whitney U (Mann & Whitney, 1947) is used. From the Test Statistics Mann-Whitney table (Table 6.7), it is inferred that in all pair of cases, there are not observed statistically significant differences ($p > 0.05$). Specifically, the greatest similarity is noted between strength and closeness and among betweenness and closeness cases ($U = 107.5$, $p = .836$). Furthermore, the 'inverse_strength' case reveals lower p-values with the observed cases ($p \leq 0.34$). Nevertheless, the differences are not significant since they are higher than $p > 0.05$.

Therefore, it is concluded that if the variables are placed in the weight matrix in different orders, the result of the V_{11} would not be a statistically significant difference when the structures of the models are hierarchical. Nevertheless, if there is a lack of expert knowledge, we suggest the researchers to rank the variables according to their strength-centrality indices to create the hierarchical structures. One evidence for in support of this suggestion can be found in network analysis as explained in Chap. 5 which refers that the most powerful concept connects with the highest direct weights can affect and influence more variables in the graph. One more evidence can be found in the results of the bootstrap method. The case-dropping bootstrap is applied to test the overall stability by eliminating gradually observations from the initial dataset (Sect. 5.4). The results from the correlation stability analysis (Table 6.1) reveal that the produced models are stable; specifically, the strength-centrality index shows higher stability.

6.5 Conclusions

Most of the FCM studies used experts' knowledge, literature or historical data to model a complex problem. Nevertheless, the fundamental challenge of this study was the learning procedure of the weights among the observed concepts directly from the available data without any intervention of experts or historical data. For this reason, we introduced a complete data-driven learning algorithm, the glassoFCM which has modelled and has simulated a job-satisfaction problem in Belgium. Specifically, the resulting FCM model was used to analyze and simulate the influence of 11 parameters and predict the behavior of the output concept to increase the willingness of the employees to work.

Since 11 nodes participated, 10 input variables and 1 the output variable, the glassoFCM algorithm was applied to estimate the undirected network structures. Then, the case-dropping bootstrap method was applied to check the stability of the models. Afterwards, the undirected graphs were transformed to directed graphs indicating the influence of the edges among concepts. In particular, the concepts (nodes) in the weight matrix were rearranged according to six rankings (strength, closeness, betweenness, mean, random and inverse_strength) and the upper triangular part of the weighted matrix was used to interact with the initial vector in the simulation process of the glassoFCM.

From this research, two important scientific findings were observed. First, the results showed that rearranging the input values and maintaining the target concept in the last position of the FCM weight matrix the values of the output concept were not changed significantly. Consequently, our approaches could be effective in problems when expert knowledge and/or historical data could not determine with high assurance the directions of the edges among concepts of an FCM. As a consequence, this result has further strengthened our confidence to make meaningful decisions/predictions in hierarchal FCM structures, when the value of a single output (or receiver) concept is examined. Second, the highest influence of the output concept 'Willingness to work' was derived when concepts of its neighborhood were activated. Therefore, in the observed job satisfaction problem, if the concepts that were connected with the output concept were improved, then the employees' willingness to work would be increased and consequently, the productivity and the commitment to the company would be increased as well.

Generally, in cases that the knowledge of an expert is not available, we recommended the researchers to place the concepts according to their strength-centrality. This could have explained for two reasons i) the most powerful concept directly had affected and has influenced more variables in the graph and ii) the results from the correlation stability analysis revealed that the produced models according to strength were in a general reliable.

References

Amirkhani, A., Papageorgiou, E. I., Mohseni, A., & Mosavi, M. R. (2017). A review of fuzzy cognitive maps in medicine: Taxonomy, methods, and applications. *Computer Methods and Programs in Biomedicine, 142*, 129–145.

Bollen, K. A. (1989). *Structural equations with latent variables* (pp. 263–269). Wiley.

Dikopoulou, Z., & Papageorgiou, E. (2017). Inference of Fuzzy Cognitive Maps (FCMs). The Comprehensive R Archive Network (CRAN), R package Version 0.1.3. https://cran.r-project.org/web/packages/fcm/vignettes/vignettes.html

Dikopoulou, Z., Papageorgiou, E., Jetter, A., & Bochtis, D. (2018). *Open-source tool in R language to estimate the inference of the Fuzzy Cognitive Map in environmental decision making*. 9th International Congress on Environmental Modeling and Software, Fort Collins, USA, June 24–28, 2018.

Dikopoulou, Z., Papageorgiou, E., Mago, V., & Vanhoof, K. (2017). A new approach using mixed graphical model for automatic design of fuzzy cognitive maps from ordinal data. *IEEE International Conference on Fuzzy Systems (FUZZ-IEEE)*. https://doi.org/10.1109/fuzz-ieee.2017.8015607

Dikopoulou, Z., Papageorgiou, E., & Vanhoof, K. (2020). *From undirected structures to directed graphical lasso fuzzy cognitive maps using ranking-based approaches*. IEEE International Conference on Fuzzy Systems (FUZZ-IEEE).

Fisher, R. A. (1918). The correlation between relatives on the supposition of Mendelian inheritance. *Philosophical Transactions of the Royal Society of Edinburgh, 52*, 399–433.

Knowland, V. C. P., Purser, H., & Thomas, M. S. C. (2015). Cross-sectional methodologies in developmental psychology. *International Encyclopedia of the Social & Behavioral Sciences*, 354–360. https://doi.org/10.1016/B978-0-08-097086-8.23235-2

Kruskal, W. H., & Wallis, W. A. (1952). Use of ranks in one-criterion variance analysis. *Journal of the American Statistical Association, 47*(260), 583–621. https://doi.org/10.1080/01621459.1952.10483441

Mann, B. H., & Whitney, R. D. (1947). On a test of whether one of two random variables is stochastically larger than the other. *Annals of Mathematical Statistics, 18*(1), 50–60.

Papageorgiou, E. I., & Salmeron, J. L. (2013). A review of fuzzy cognitive maps research during the last decade. *Fuzzy Systems, IEEE Transactions, 21*, 66–79.

Papageorgiou, E. I. (2014). Fuzzy cognitive maps for applied sciences and engineering from fundamentals to extensions and learning algorithms. *Intelligent Systems Reference Library, 54*.

Schneider, M., Shnaider, E., Kandel, A., & Chew, G. (1998). Automatic construction of FCMs. *Fuzzy Sets and Systems, 93*(2), 161–172.

Shapiro, S. S., & Wilk, M. B. (1965). An analysis of variance test for normality (complete samples). *Biometrika, 52*(3–4), 591–611.

Stach, W., Pedrycz, W., & Kurgan, L. A. (2012). Learning of fuzzy cognitive maps using density estimate. *IEEE Transactions on Systems, Man, and Cybernetics. Part B, 42*(3), 900–912.

Tukey, J. W. (1949). Comparing individual means in the analysis of variance. *Biometrics., 5*(2), 99–114.

Chapter 7
Thesis Conclusions

This chapter summarizes the most important achievements of the proposed data-driven algorithm, the glassoFCM such as the interpretability of the produced model reducing the complexity of the graph, the identification of the relevant variables that affect significantly the output value and high performance in term of time execution comparing to a distance-based method. Moreover, the research conclusions, the contribution to knowledge, the recommendation for practitioners, the limitation of the glassoFCM algorithm and the future results is described.

7.1 Introduction

To sum up, this dissertation provided an overview of modeling, testing, analyzing and simulating cross-sectional data in the business area. Particularly, 56,246 employees' preferences concerning the job-satisfaction problem in Belgium were evaluated on a five-Likert scale ranging from 1 (Strongly disagree) to 5 (Strongly agree).

According to the literature review, most of the studies have been used expert knowledge or methods to handle datasets to model various problems. The major limitation of *human-driven models* is the intervention of experts to indicate the strength and the causal-effect relationship (the direction) of the edges. One the other hand, *data-driven methods* (such as Factor Analysis), estimate latent structures in which unobservable latent variables share their variance with a set of items. Nevertheless, not all real complex problems could be explained using latent structures; particularly, when they do not present similar properties. Consequently, it is tough to classify them into different groups. Some problems are more appropriate to be represented as networks in which the direct relations between the observed variables are reflected.

For this reason, the distance-based method (Schneider et al., 1998) was applied in this research to indicate the weight matrix which was emerged by calculating the

Z. Dikopoulou, *Modeling and Simulating Complex Business Perceptions*, Fuzzy Management Methods, https://doi.org/10.1007/978-3-030-81496-0_7

distances of the vectors (variables). However, some parameters of the matrix were misleading since they revealed complete relations (all variables were fully connected, $\delta = 1$). Therefore, the interpretability of the model was decreased and expert knowledge was required to determine the causality among variables. Generally, the distance-based method is a merely data-driven (semi-automated) approach.

Consequently, the proposed methodology, the graphical lasso with EBIC regularization was deemed as an efficient data-driven technique, capable to model cross-sectional ordering data, eliminating the false positives and false-negative connections and derive "*decipherable*" (sparser) networks. Next sections of this final chapter will provide: i) *research conclusions,* a summary of our findings answering the research questions of Chap. 1, ii) *contribution to knowledge* highlighting how this research has effectively conducted to "fill the gap" of previous studies that were questionable or problematic (as they were referred in the introduction of this dissertation), iii) *recommendations* to the readers based on current conclusions and future suggestions, iv) *limitations* of the methods of this dissertation and v) finally, *future research,* includes several future research suggestions.

7.2 Research Conclusions

As stated in the Introduction, the dissertation was conducted to identify effective methods to model, test, analyze, simulate cross-datasets and determine the most important variables that influence the output variable V_{11} ("Willingness to work"). Below, the resulting conclusions of this research are listed.

i) *Data modeling* (Chap. 4). A method from graphical models, the glasso with EBIC regularization was deduced as a suitable method for modeling network structures straight from ordinal datasets. The learning process of the weight edges is conducted without the usage of historical data or knowledge about the associations among variables. Moreover, the glasso algorithm returned a symmetric sparser weight matrix since small partial correlation coefficients were shrunk exactly to zero in which spurious edges were omitted from the model. Importantly, our results provide evidence for sparsity in 15 network structures estimated with glasso method (0.699 ± 0.111) comparing to the distance-based method (1.000 ± 0.000). To our knowledge, we first applied the graphical lasso approach to model a business problem.
ii) *Checking for efficient models* (Chap. 5). Two bootstrap methods have utilized, the *sampling with replacement* and the *case-dropping subset* which were used to test the *accuracy* and the *stability* of each estimated network, respectively. The accuracy of the network has occurred from the differences between the mean of the bootstrap samples and the actual weights of the estimated graph. The results of the accuracy point that Energy Services (S_{10}), Food (S_4) and Automotive (S_1) sectors obtained higher differences. Nevertheless, stability testing was important to validate the results of accuracy. In this bootstrap method, the sample size was

minimized from 95% to 5% comparing to the initial dataset. In this samples, the correlation stability coefficient was checking the modifications of the centrality indices. Sectors S_{10} and S_1 were eliminated from the simulation process since the CS-coefficients were lower than 0.5.

iii) Searching for potential clusters (Chaps. 4 and 5). For the formation of any group of variables, we performed two analyses; the first was applied in the dataset (a-priori analysis) and the other in the model (a post-hoc analysis). In the a-priori analysis, nine different routines were implemented to investigate the proper number of clusters that each dataset suggested. As a result, the majority of the methods (67%) for the working sectors proclaimed that the cumulative variance of the variables could be explained with only one cluster. Consequently, it was inferred that the observed job-satisfaction problem could be efficiently represented as a network rather than using any dimensional reduction method. After estimating a network, a post-hoc analysis was performed to probe if some nodes of the graph tended to be or not be clustered to each other. In other words, we were investigated if the structures of job satisfaction networks obtained small-world characteristics. For this reason, various graph indices were estimated such as clustering coefficient, transitivity, average length and 'small-worldness' (*SW*) property. To recall, the small-worldness is defined by the ratio of the standardized transitivity value to the standardize average distance between two nodes (Humphries and Gumey, 2008). The small-worldness values of the job-attractiveness networks were indicating that none of the networks were shown small-world properties. Consequently, they were classified as random networks since the SW values were below three. Therefore, two conclusions were inferred; first, the nodes were not significantly clustering together and second, there was no possible prediction for the connections of a new node in the network since the important properties of random graphs could be appeared quite suddenly (Erdős & Rényi, 1959).

iv) *Simulation process* (Chaps. 4 and 6). Since glasso method provides symmetric weighted graphs and there was a lack of expert knowledge to define the direction of the edges, it was impossible to simulate the network. Therefore, we included in the initial dataset, one output variable the V_{11} (Willingness to work) to investigate which from the input variables (V_1–V_{10}) could affect more the V_{11}. The procedures of modeling and bootstrapping were applied again. For the simulation process, the Fuzzy Cognitive Map (FCM) methodology was used to predict the output value of V_{11} by activating a subset of the input variables. In order to simulate the influence of the input variables (V_1–V_{10}) on the output variable (V_{11}), the upper triangular part of the weighted symmetric matrix was obtained by placing the V_{11} in the last position. Moreover, for the sake of completeness, we developed in the R programming language and presented a step-by-step guide of the '*fcm*' package. The main function of the '*fcm*', the *fcm.infer*() operates eight input arguments the initial vector, the weighted matrix, the number of iterations, the inference rule, the threshold function, the lambda and the epsilon parameter. As a result, the fcm.infer() function returns three objects:

(a) a data frame with the values of the concepts in every iteration, (b) an FCM map and (c) the convergence plot of κ steps.

v) *Identifying the important variables* (Chap. 6). According to the collaborating company, *the most significant variables of a dataset were these with a higher sum score*. However, the main weakness of the perception was the violation of the ordinal type of data since it was falsely assumed that the numerical labels ('1', '2', '3', ...,) could be indicated with equal intervals between levels. The evidence of the evaluation of the most critical variables in the model was presented in Chap. 6. Particularly, after 15 what-if scenarios and six rearrangements of the input variables that were occurred for each sector, the output value (V_{11}—Willingness of the employees to work) of the FCM was significantly affected when the direct and positive connected variables with V_{11} were activated. To validate the results of the FCM simulation, the variables of the initial cross-sectional dataset were filtered with the highest values of the Likert scale ('5'). This procedure was repeated twice for the neighbors of V_{11} and for the variables with the highest mean value. The result showed that the output variable V_{11} was obtained higher value when the neighbors of V_{11} were filtered with the value 5.

vi) *Implementing the 'fcm' R package* (Chap. 4): We developed an open-source R package, the *fcm* (Dikopoulou & Papageorgiou, 2017) to accomplish the FCM inference of the system using seven input arguments in the *fcm.infer*() function. Moreover, it provides six different inference rules (Kosko, modified-Kosko, rescale, clamped-Kosko, clamped-modified-Kosko and clamped-rescale) and four threshold functions (bivalent, trivalent, sigmoid and hyperbolic tangent) to calculate the FCM inference. Finally, the fcm.infer() function return three objects: (a) the values of the concepts in every iteration, (b) the FCM map and (c) the converged plot of κ simulation steps.

7.3 Contribution to Knowledge

Generally, research is directed at triggering, supporting or improving existed methods or/and developing new methods to enhance the knowledge and understanding for reconsidering the way we observe our problem. This dissertation highlights the significance of research with the following points:

i) *First contribution to the business area* (Chap. 4): Until today, the regularized network estimation has been applied widely to a psychological field (e.g., De Ron et al., 2019; Isvoranu et al., 2019; Epskamp & Fried, 2017; Epskamp et al., 2017a, b; Fried et al., 2016; McNally, 2016; Costantini et al., 2015; van Borkulo et al., 2015; Kossakowski et al., 2015). However, the graphical lasso with EBIC regularization was not yet introduced in business literature. For this reason, we performed this methodology at cross-sectional ordinal data to model a job-satisfaction problem.

ii) *Proposing the combined glassoFCM method* (Chaps. 3, 4 and 6): The aim of our research was not only to introduce of an automated data-driven learning procedure of the weights among the observed concepts but also to simulate different scenario policies for decision making according to the results of the output value. Therefore, we proposed the *glassoFCM* method which was a compound of the graphical lasso with EBIC regularization and the FCM. To conclude, the main objective of this chapter was the introduction of the glassoFCM to model and simulate a job-satisfaction problem in Belgium.

iii) *Confirming the current knowledge* (Chaps. 5 and 6): Our findings confirm the previous results of Epskamp, Borsboom and Fried (Epskamp et al., 2017a) that large datasets provided more consistent models. Evidence for our results could be found in Chap. 5 which the accuracy and stability of the models were estimated. In particular, the non-parametric bootstrap (resampled with replacement) revealed that the sum of the absolute differences (SAD) among the sample weights and the mean bootstrap weights (Table 5.4) was lower in larger datasets (such as S_8 and S_{15}) and higher in smaller datasets (such as S_{10}). Moreover, the case-dropping subset bootstrap for stability testing (Table 5.5) had shown that the CS-coefficient of centrality indices was higher (close to 0.75) in larger sample sizes (S_8 and S_{15}) and lower than 0.5 in lower sample sizes (S_1 and S_{10}). Furthermore, including an additional variable in the datasets, the glasso algorithm estimated more stable models. Specifically, the CS-coefficient of centrality indices of the job satisfaction models in Table 6.1 (networks with 11 variables) determined higher values compared to Table 5.5 (networks with 10 variables).

iv) *Evaluating distance-based models in large datasets* (Chap. 4): According to the literature review, one way to create FCMs was by using experts' knowledge or theory, known as *expert-based models*. Another way was to used *semi-automated data-driven approaches* such as the distance-based method (Schneider et al., 1998) to estimate symmetrical weight matrices directly from datasets. Nevertheless, after completing the matrix of SIRM, some data could be misleading because some vectors were unexpectedly related to each other providing dense network structures (Schneider et al., 1998; Rodriguez-Repiso et al., 2007; Arvan et al., 2016; Bakhtavar & Shirvand, 2018; Zarrin & Azadeh, 2018; Stylios et al., 2020). For this reason, an expert intervention was required to remove edges from the model and indicate the causal-effect relationships (directions among edges). Moreover, we first evaluated and reported the failure of the distance-based method to model large datasets (Dikopoulou et al., 2017). The reason was the high calculated weights between the concepts and the produced results from the scenario analysis since there were not observed deviations on the simulated values of the output concept for the conducted scenarios. Therefore, it was unreasonable to make decisions and policies using large datasets.

v) *Simulating fully automated data-driven FCMs with the glassoFCM method* (Chap. 6): The main limitation of the FCM approach was the significant dependency on the knowledge and experience of domain experts (Salmeron, 2010). For this reason, this dissertation proposed the *glassoFCM* method to overcome this problem since it modeled large sample sizes directly from datasets and then, it simulated various scenarios. First, the symmetric weighted matrix (which learned from glasso method) was converted into an asymmetric weighted matrix. Consequently, sparser graph neural networks were estimated with a single output (V_{11}). Validating the value of the target concept (V_{11}) by rearranging[1] the input variables in the upper triangular part of the weighted matrix. This process was useful to investigate if the output values were significantly changed in the scenario analysis. Moreover, another contribution in these structures was the identification of the higher value of V_{11} according to the rearrangements. The results had indicated that the maximum value of V_{11} was obtained when the input variables were ordered according to strength-centrality ranking (Dikopoulou et al., 2020a, b).

vi) *Indicating the most important variables of the model with a single output* (Chap. 6): As referred in Sect. 7.2, the most important variables in the particular network structure (graph neural network with a single output) were identified. Thus, it was studied that the variables which were directly connected to the output concept were the most significant. The evidence of these results was derived from the FCM simulations which were validated using six observed cases of 15 different what-if scenarios. Our findings (Table 6.3) indicated that the V_{11} were obtained the higher value (14th scenario) when its neighbors were activated in the scenario analysis. Moreover, Table 6.4 presented the comparisons of two activated scenarios of three working sectors. Specifically, the first initial vector activated the neighbors of the V_{11} and the second had activated the variables with the highest mean score. The results showed that the maximum values of V_{11} were obtained when its neighbourhood was activated (Dikopoulou et al., 2020b).

vii) *Developing the 'fcm' R package* (Chap. 3): Since the FCM methodology already was used in a substantive number of publications in diverse fields (Dikopoulou et al., 2020a; Papageorgiou & Salmeron, 2013), we implemented the FCM methodology in the free (open source) *'fcm'* R package (Dikopoulou et al., 2018; Dikopoulou & Papageorgiou, 2017) which is available in CRAN repository (*https://cran.r-project.org/web/packages/fcm/*). Thus, *the 'fcm'* package is designed to be usable by researchers that are familiar with R or beginners with no experience in any programming language. Moreover, we presented a tutorial on simulating FCMs accompanied by R syntax.

Particularly, the *fcm.infer*() function was accomplished the FCM simulation utilizing seven input arguments: the activation vector, the weighted matrix, the number of iterations, the inference rule, the threshold function, the value for the parameters λ and ε. The algorithm of the *fcm.infer* function stopped when the equilibrium point for each concept was reached or the number of iterations was completed. Finally, the fcm.infer() function returned three objects: (a) the values of

[1] The concepts in the upper triangular part of the weight matrix were rearranged according to six rankings (strength, closeness, betweenness, mean, random and inverse_strength).

the concepts in every iteration, (b) the FCM map and (c) the converged plot of κ simulation steps. To conclude, the 'fcm' attracts the attention of the researchers since more than 15,462 downloads have been transferred from the RStudio CRAN Mirror (https://gertjanssenswillen.shinyapps.io/downloads_cran_edear/).

In total, this study provides the framework for a new way to model and simulate cross-sectional ordinal datasets.

7.4 Recommendations for Practitioners

The dissertation has shown a data-driven approach to model and has simulated cross-sectional ordinal datasets. Below, recommendations for experts (researchers or non-researchers) are noted:

i) *Estimate semi-automated structures*. In cases that expert knowledge is available, experts can distinguish the causal-effect relationships among the concepts in the symmetric weight matrix deriving from the glasso (with EBIC regularization) algorithm (Fig. 7.1). Therefore, a graph-based structure will be estimated. Furthermore, in the simulation process, the values of all concepts can be evaluated simultaneously (when the equilibrium point of the system will be reached); not only from one output concept.
ii) *For glasso models use the clamped inference rules*. The glasso method usually estimates sparser models since the algorithm removes spurious connections from the model. Thus, it returns (normally) a matrix in which many partial correlation weights are low. If the estimated weights are above 0.8, this could indicate near-perfect collinearity between variables (Epskamp & Fried, 2017). These models could be revealed when the sample size is very small for the number of nodes in the network or the estimation of the polychoric correlation matrix is not positive-definite (all eigenvalues are not higher than zero). Therefore, for better results of the FCM simulation process, we recommend the users to select the rescale inference rule (such as the rescale-clamped '*rc*'), since, it boosts in every simulation step the subset of the activated variables.

Consequently, the interpretation of the results is simpler because the values of the concepts present high variance and it is easier to identify which concept of the FCM was influenced more.

7.5 Limitations

We are aware that our research may have five limitations focusing on the estimated regularized networks as described in the previous study (Epskamp & Fried, 2017) and the FCM simulation procedure. Below, the weaknesses of the glasso and the FCM methodology will be discussed:

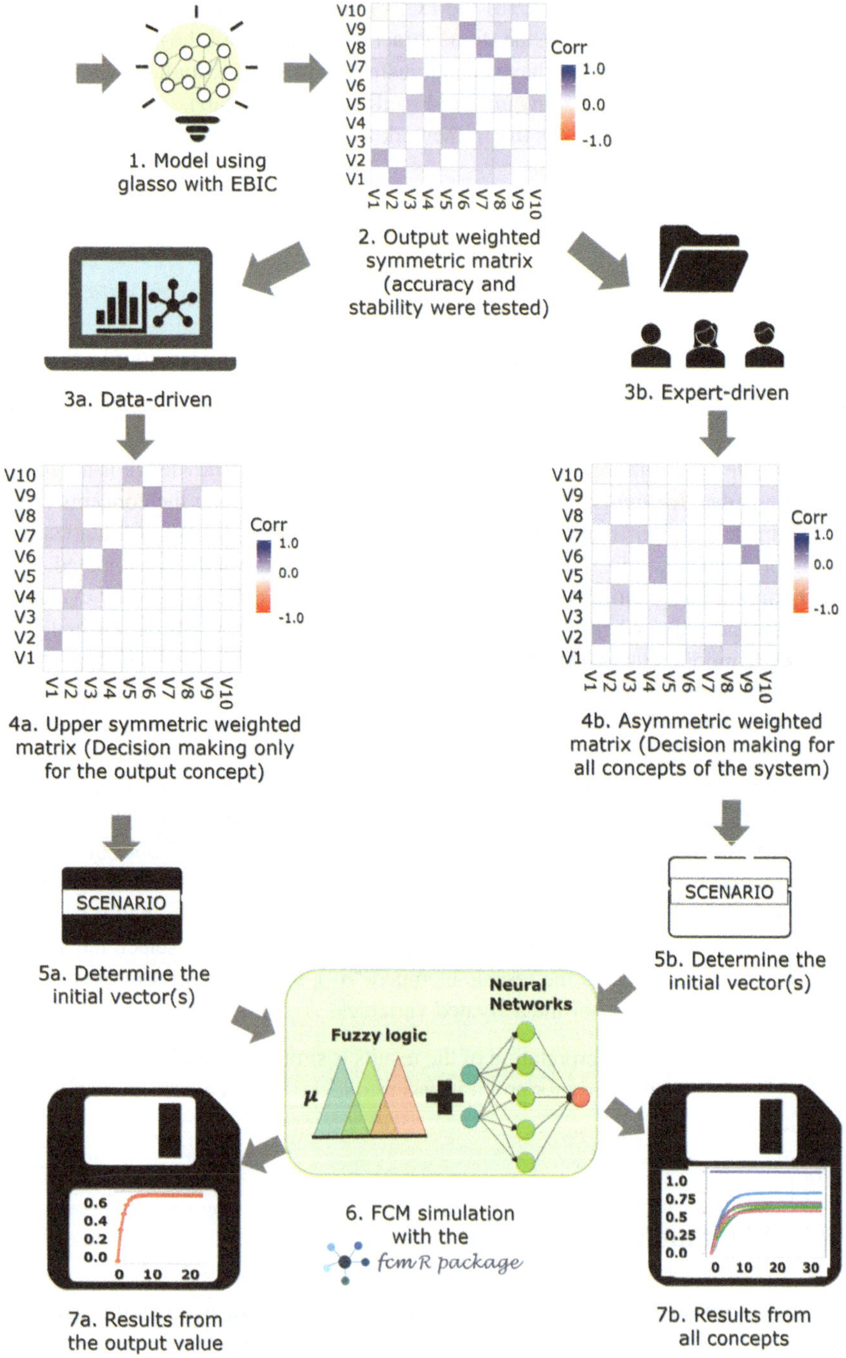

Fig. 7.1 The simulation of *fully-automated* network structures (data-driven model; left) and the simulation of *semi-automated* graph structures (expert-based model; right)

i) *The estimated network has no or low connectivity*. In such occasions, the partial connections between the variables cannot be inferred due to the very low sample size in the network. As referred in the Introduction of this thesis, data-driven models require a sufficient amount of training data. Moreover, the EBIC penalizes the edge weights to minimize false-positive associations. This indicates that the partial correlation matrix gradually is getting similar to the regularized partial correlation matrix as the sample size increases. One solution to address this problem is to set the hyperparameter gamma lower ($\gamma = 0$). Nevertheless, this adjustment induces the increase of spurious connections. Another alternative is to estimate a robust network from a subset of variables. Selecting a number of the most significant variables, less but accurate edges will be estimated since will be more observations per parameter.

ii) *The estimated network has very high connectivity*. The phrase "very high connectivity" refers to networks that the nodes are densely linked including unexpected negative connections and many unreasonably high partial correlations (near -1 and 1) indicating almost perfect collinearity among nodes. In such cases, the network structure should not be interpreted since lasso returns sparser networks with a lower partial correlation between variables. This occurs when the input correlation matrix is not definite positive due to small sample size or mistaken polychoric correlations. A previous study (Epskamp & Fried, 2017) suggested to compare networks from polychoric correlation with networks from Spearman correlation; the obtained networks must be similar. If the resulting correlations differ significantly, then the polychoric correlations may not be trustworthy.

iii) *The estimated network is relatively sparse, but some edges among variables are extremely high*. This occurs when the sample size is small; therefore, the cells of the pairwise cross-tables of the variables that are necessary to indicate the polychoric correlations could be zero. This could determine unstable estimated polychoric correlations and unreliable partial correlations. The solution to this problem is connected with the previous problem. The networks derived from polychoric and Spearman correlations should be compared.

iv) *More concepts, a more complex simulation process*. We presented that the stability of the estimated models with glasso was improved when an additional variable was included (Table 6.1). However, increasing the number of concepts in the FCM has a negative impact on the simulation time (the execution time will be increased as well). Moreover, an FCM with many concepts is difficult to be analyzed and interpreted (Poczęta et al., 2017). Therefore, low-dimensional data can be simulated if the accuracy and stability of the estimated networks are acceptable.

v) *Data-driven FCMs in a specific network structure*. This dissertation has tended to focus on a data-driven approach. However, this approach was effective in simpler (graph-neural) structures with a single output concept rather than an abstract graphical structure. For this reason, it is possible to make about the output concept and not for all concepts.

As it is observed, our work clearly has some limitations. Nevertheless, we believe that this research could be the basis for data-driven FCMs.

7.6 Future Research

Many different adjustments, examinations and simulations have been left for the future. The main reason is the time since the experiments with real data are usually time-consuming. Therefore, future work involves deeper analysis of particular methods, a suggestion of advanced methods or simply interest for further research. The following future objectives could be investigated:

i) *Apply the glassoFCM to more groups*. Since our results are encouraging, we are in the process of investigating the impact of the job-satisfaction variables to the employees' willingness to work in different categories such as gender (male and female), education (primary/secondary, bachelor, master), age (18–24, 25–44, 44–65), occupation (employed or unemployed), activity (product worker, office worker and manager), countries (Canada, Argentina, USA, Australia, China, Japan, UK, Germany, Italy, Switzerland, etc.).
ii) *Increase the number of variables*. As discussed in the previous subchapter, in large datasets if the number of variables increase, then more robust structures will be created. Therefore, to further our research, we intend to investigate how the modeling and the simulation results will be affected if we include additional variables. More variables will not only be correlated to job satisfaction but also employee's psychology (for instance '*Job stress*').
iii) *Identify automatically direct edges among variables*. Future work will concentrate on a method that will be able to identify the directions of the edges from the data. Then, graph structures will be estimated. Consequently, the FCM simulation process will be performed focusing not only on a single output concept but on all concepts of the system.
iv) *Apply glassoFCM to various types of data*. In future investigations, it might be possible to use the glassoFCM algorithm for modeling and simulating network structures from different types of data such as binary, continuous or mixed data.
v) *Develop a Shiny application*. Shiny (*https://shiny.rstudio.com*) is an R package that creates interactive web apps straight from R. Moreover, it is a great tool for prototyping since it could display simple tables and figures to a fully interactive dashboard. Therefore, a shiny app will permit the decision-makers to get insights from data (analyzing, visualizing, modeling and simulating) without coding in R.

References

Arvan, M., Omidvar, A., & Ghodsi, R. (2016). Intellectual capital evaluation using fuzzy cognitive maps: A scenario-based development planning. *Expert Systems with Applications, 55*, 21–36.

Bakhtavar, E., & Shirvand, Y. (2018). Designing a fuzzy cognitive map to evaluate drilling and blasting problems of the tunneling projects in Iran. *Engineering with Computers*. https://doi.org/10.1007/s00366-018-0581-y

Costantini, G., Epskamp, S., Borsboom, D., Perugini, M., Mõttus, R., Waldorp, L. J., & Cramer, A. O. J. (2015). State of the aRt personality research: A tutorial on network analysis of personality data in R. *Journal of Research in Personality, 54*, 13–29.

De Ron, J., Fried, E. I., & Epskamp, S. (2019). Psychological networks in clinical populations: Investigating the consequences of Berkson's bias. *Psychological Medicine*.

Dikopoulou, Z., & Papageorgiou, E. (2017). *Inference of Fuzzy Cognitive Maps (FCMs)*. The Comprehensive R Archive Network (CRAN), R package Version 0.1.3. https://cran.r-project.org/web/packages/fcm/vignettes/vignettes.html

Dikopoulou, Z., Papageorgiou, E., Jetter, A., & Bochtis, D. (2018). *Open-source tool in R language to estimate the inference of the Fuzzy Cognitive Map in environmental decision making*. 9th International Congress on Environmental Modeling and Software, Fort Collins, USA 24–28, 2018.

Dikopoulou, Z., Papageorgiou, E., Mago, V., & Vanhoof, K. (2017). A new approach using mixed graphical model for automatic design of fuzzy cognitive maps from ordinal data. *IEEE International Conference on Fuzzy Systems (FUZZ-IEEE)*. https://doi.org/10.1109/fuzz-ieee.2017.8015607

Dikopoulou, Z., Papageorgiou, E., & Vanhoof, K. (2020a). Retrieving sparser fuzzy cognitive maps directly from categorical ordinal dataset using graphical lasso models and the max-threshold algorithm. IEEE International Conference on Fuzzy Systems (FUZZ-IEEE).

Dikopoulou, Z., Papageorgiou, E., & Vanhoof, K. (2020b). From undirected structures to directed graphical lasso fuzzy cognitive maps using ranking-based approaches. IEEE International Conference on Fuzzy Systems (FUZZ-IEEE).

Epskamp, S., & Fried, E. I. (2017). A tutorial on regularized partial correlation networks. *Psychological Methods*. In press. https://doi.org/10.1037/met0000167

Epskamp, S., Borsboom, D., & Fried, E. I. (2017a). Estimating psychological networks and their accuracy: A tutorial paper. *Behaviour Research Methods, 2017*, 1–18.

Epskamp, S., Kruis, J., & Marsman, M. (2017b). Estimating psychopathological networks: Be careful what you wish for. *PLoS One, 12*(6), e0179891. https://doi.org/10.1371/journal.pone.0179891

Erdős, P., & Rényi, A. (1959). On random graphs. I (PDF). *Publicationes Mathematicae, 6*, 290–297.

Fried, E. I., Epskamp, S., Nesse, R. M., Tuerlinckx, F., & Borsboom, D. (2016). What are 'good' depression symptoms? Comparing the centrality of DSM and non-DSM symptoms of depression in a network analysis. *Journal of Affective Disorders, 189*, 314–320.

Humphries, M. D., & Gurney, K. (2008). Network 'small-world-ness': A quantitative method for determining canonical network equivalence. *PLoS One, 3*(4), e0002051. https://doi.org/10.1371/journal.pone.0002051

Isvoranu, A. M., Guloksuz, S., Epskamp, S., van Os, J., Borsboom, D., & GROUP. (2019). Towards incorporating genetic risk scores into symptom networks of psychosis. *Psychological Medicine*.

Kossakowski, J. J., Epskamp, S., Kieffer, J. M., van Borkulo, C. D., Rhemtulla, M., & Borsboom, D. (2015). The application of a network approach to health-related quality of life (HRQoL): Introducing a new method for assessing hrqol in healthy adults and cancer patient. *Quality of Life Research, 25*, 781–792.

McNally, R. J. (2016). Can network analysis transform psychopathology? *Behaviour Research and Therapy, 86*, 95–104. https://doi.org/10.1016/j.brat.2016.06.006

Poczęta, K., Kubuś, Ł., & Yastrebov, A. (2017). Concepts selection in fuzzy cognitive map using evolutionary learning algorithm based on graph theory metrics. In M. Ganzha, L. Maciaszek, & M. Paprzycki (Eds.), *Communication Papers of the 2017 Federated Conference on Computer Science and Information Systems. ACSIS* (Vol. 13, pp. 89–94).

Rodriguez-Repiso, L., Setchi, R., & Salmeron, J. L. (2007). Modeling IT projects success with fuzzy cognitive maps. *Expert Systems with Applications, 32*(2), 543–559.

Salmeron, J. L. (2010). Fuzzy cognitive maps-based IT projects risks scenarios. In M. Glykas (Ed.), *Fuzzy cognitive maps* (Studies in fuzziness and soft computing) (Vol. 247). Springer.

Schneider, M., Shnaider, E., Kandel, A., & Chew, G. (1998). Automatic construction of FCMs. *Fuzzy Sets and Systems, 93*(2), 161–172.

Stylios, C. D., Georgopoulos, B. E., & V. C. (2020). Impact and applications of fuzzy cognitive map methodologies. In O. Kosheleva, S. P. Shary, G. Xiang, & R. Zapatrin (Eds.), *Beyond traditional probabilistic data processing techniques: Interval, fuzzy etc. Methods and their applications* (Studies in computational intelligence) (pp. 229–246). Springer.

van Borkulo, C. D., Boschloo, L., Borsboom, D., Penninx, B. W. J. H., Waldorp, L. J., & Schoevers, R. A. (2015). Association of symptom network structure with the course of depression. *JAMA Psychiatry, 72*(12), 1219–1226.

Zarrin, M., & Azadeh, A. (2018). Mapping the influences of resilience engineering on health, safety, and environment and ergonomics management system by using Z-number cognitive map. *Human Factors and Ergonomics in Manufacturing & Service Industries*. https://doi.org/10.1002/hfm.20766

Appendix A

Z. Dikopoulou, *Modeling and Simulating Complex Business Perceptions*, Fuzzy Management Methods, https://doi.org/10.1007/978-3-030-81496-0

Table A.1 Demographic information about the number of observations, gender, education level and working activity of 15 sectors

Demographics	Automotive	Constructor	Finance	Food	Human Resources	Industrial Manufacturing	Informatics Consulting	Non-profit	Pharmaceutical Chemical	Energy (Power & Utilities &	Retail	Business Services Media	Transportation Logistics Wholesalers	Travel Leisure Hospitality	Other
N	1043	2301	3819	1655	1499	3053	3262	20378	2363	172	1469	3134	2833	1598	7667
Gender															
Male	70.7%	73.2%	50.4%	51.4%	31.7%	71.2%	78.7%	46.5%	61.0%	34.9%	59.0%	48.6%	67.4%	54.3%	47.5%
Female	29.3%	26.8%	49.6%	48.6%	68.3%	28.8%	21.3%	53.5%	39.0%	65.1%	41.0%	51.4%	32.6%	45.7%	52.5%
Ages															
18-24	10.3%	9.9%	11.0%	17.8%	17.1%	9.4%	8.5%	6.6%	4.7%	16.3%	9.0%	9.7%	10.2%	18.8%	11.6%
25-44	64.4%	67.5%	59.7%	63.3%	64.8%	60.7%	70.0%	61.1%	64.6%	60.5%	59.3%	60.7%	63.7%	50.1%	52.9%
45-65	25.3%	22.6%	29.3%	18.9%	18.0%	29.9%	21.5%	32.3%	30.7%	23.3%	31.7%	29.6%	26.1%	31.2%	35.5%
Education															
Secondary	44.7%	48.6%	16.4%	54.0%	17.0%	39.1%	19.3%	24.1%	25.4%	44.8%	50.0%	37.4%	49.0%	40.0%	37.5%
Bachelor	37.7%	26.6%	50.0%	30.3%	44.1%	39.0%	39.9%	47.0%	38.3%	50.0%	36.1%	31.1%	37.7%	49.0%	39.1%
Master	17.0%	23.7%	32.5%	13.1%	38.0%	21.2%	40.0%	27.3%	34.4%	5.2%	13.9%	30.4%	11.8%	9.5%	19.6%
N/A	0.7%	1.1%	1.2%	2.6%	0.9%	0.7%	0.7%	1.6%	1.9%	0.0%	0.0%	1.1%	1.6%	1.5%	3.8%
Activity															
Production	30.7%	28.8%	0.2%	28.9%	0.5%	35.4%	0.4%	2.6%	17.4%	7.0%	5.2%	20.6%	25.2%	26.8%	10.0%
Office	57.9%	50.4%	69.0%	56.7%	74.0%	43.4%	79.2%	80.1%	61.9%	93.0%	69.8%	58.2%	64.3%	59.1%	68.6%
Manager	11.4%	11.8%	26.1%	9.1%	20.8%	18.4%	14.4%	8.6%	17.9%	0.0%	16.6%	10.1%	9.3%	9.6%	9.2%
N/A	0.0%	9.0%	4.7%	5.4%	4.7%	2.8%	6.0%	8.7%	2.8%	0.0%	8.3%	11.1%	1.2%	4.6%	12.3%

Appendix B

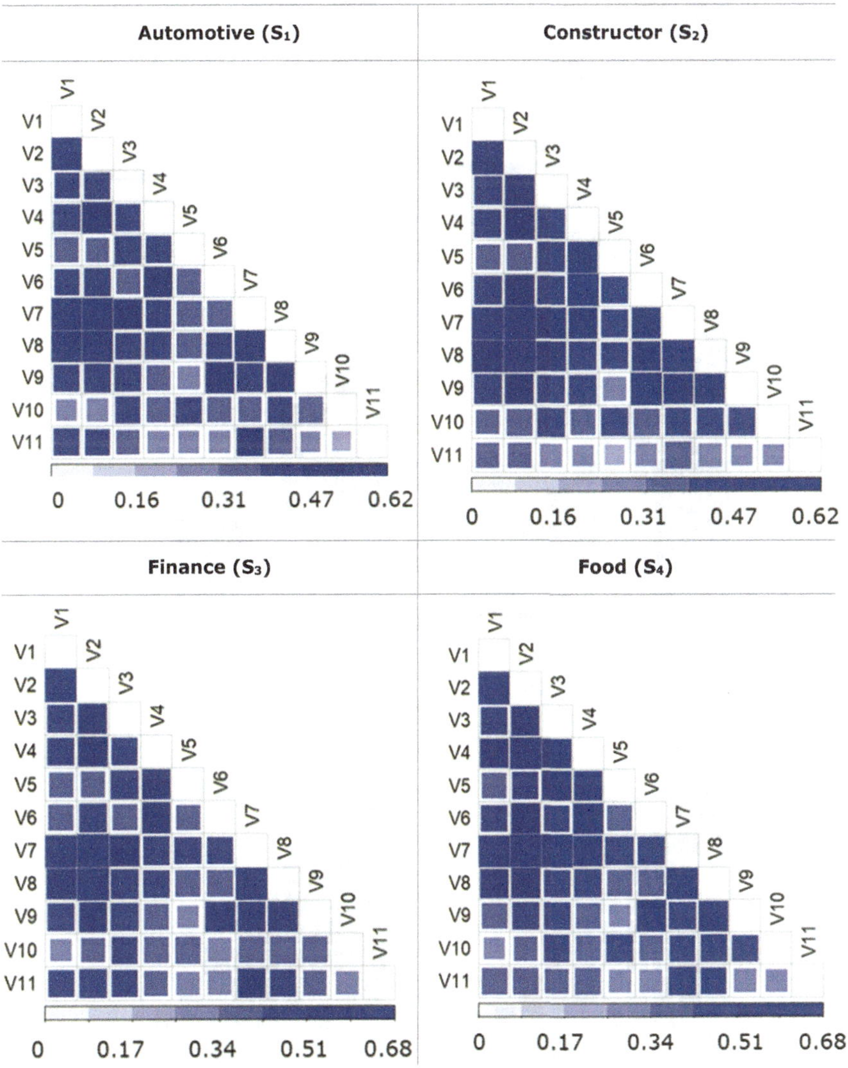

Fig. B.1 Fifteen corrplots that represent the Spearman's correlations among ten variables in each working sector. Blue color signifies positive correlations among variables

Z. Dikopoulou, *Modeling and Simulating Complex Business Perceptions*, Fuzzy Management Methods, https://doi.org/10.1007/978-3-030-81496-0

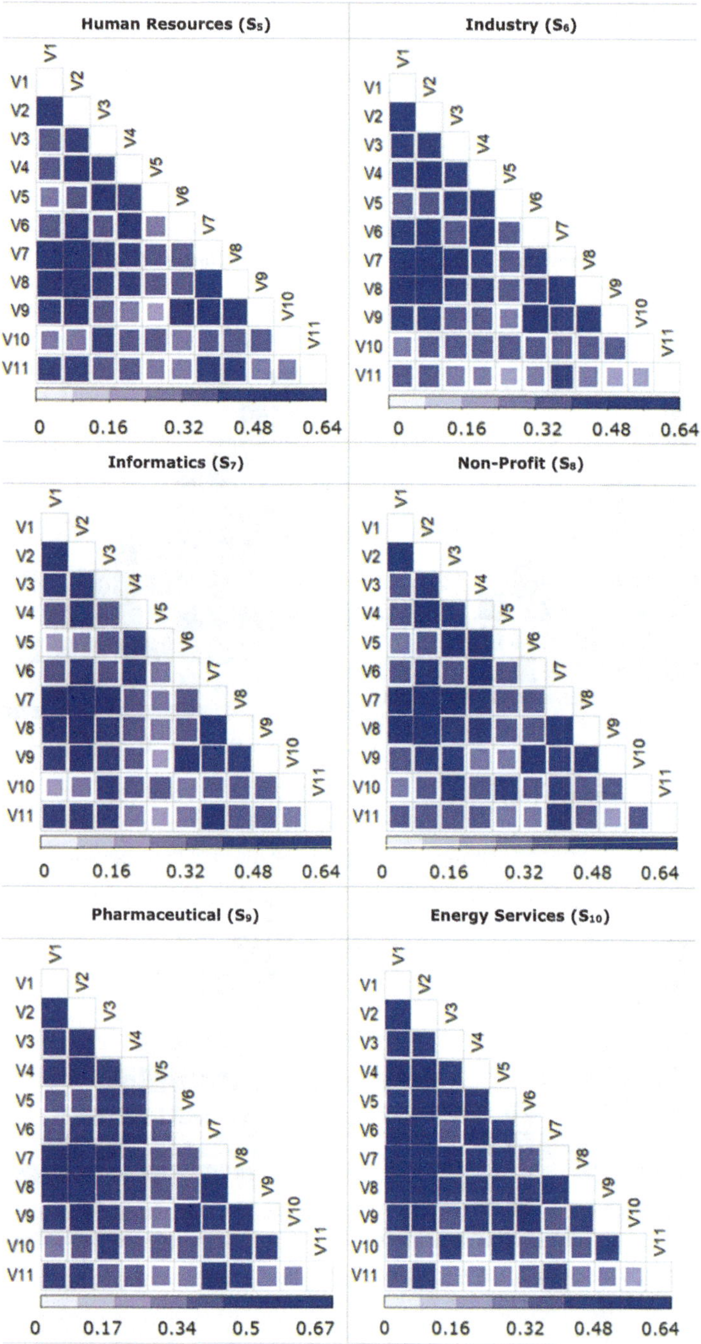

Fig. B.1 (continued)

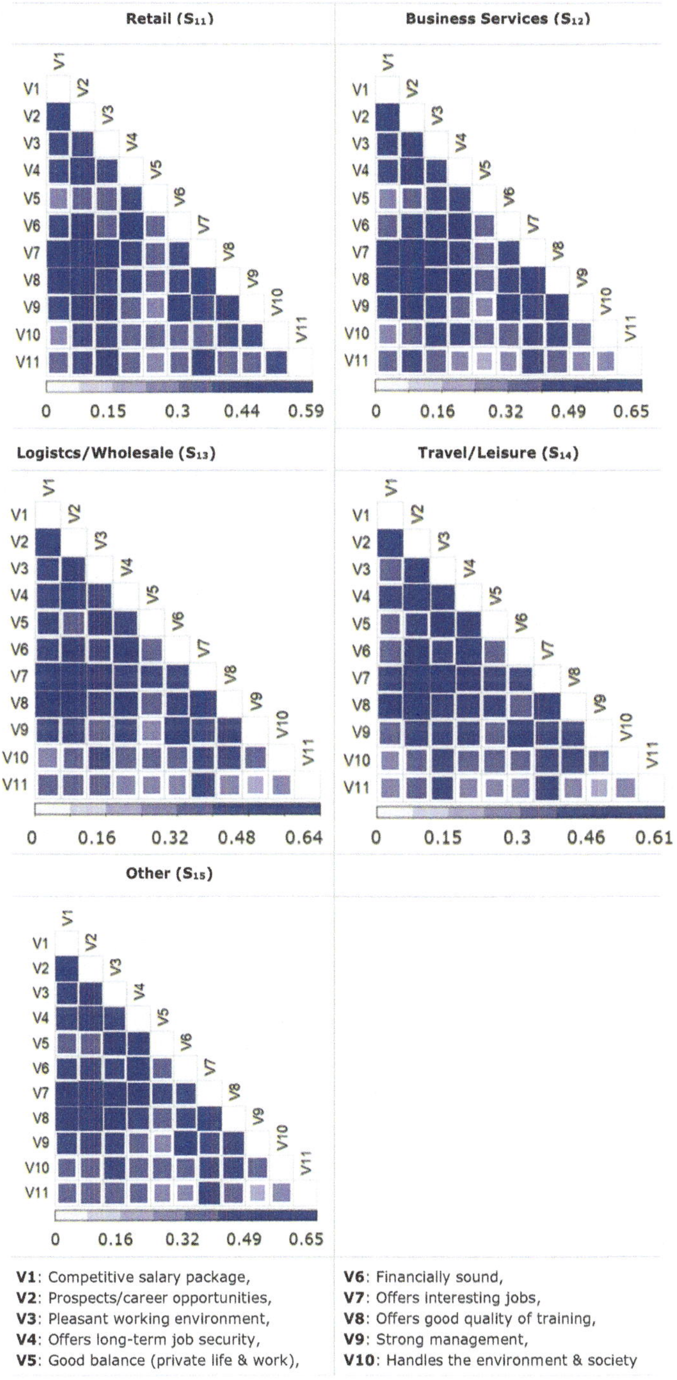

Fig. B.1 (continued)

Appendix C

The derived job-satisfaction network from the Business sector (Table C.1) is defined as a complete graph since each node is jointed to all other nodes in the network. As a result, the density is equal to one ($\delta = 1$). Therefore, high connectivity increases the number of spurious connections and leads to complex and unexplainable models. Moreover, the interpretability of the values of the output node V_{11} after the FCM simulation cannot be clarified since the output values in all scenarios are very high. Table C.2 illustrates the FCM simulations of fifteen policy-scenarios. As it is observed, the output values of V_{11} are close to one ($\mu = .981$, *st. dev* $= .039$) and they are the same in all cases (Strength, Closeness, Betweenness, Mean, Random and Reverse_strength) since the nodes are connected to each other and therefore, any rearrangement of the nodes returns the same outcome of the V_{11}.

Z. Dikopoulou, *Modeling and Simulating Complex Business Perceptions*, Fuzzy Management Methods, https://doi.org/10.1007/978-3-030-81496-0

Table C.1 The SIRM matrix (distance-based method) of 11 variables correlated to a job satisfaction problem of the Business sector (S_{12}) dataset. The first ten variables are considered as input variables (V_1–V_{10}) and the last one as the output variable (V_{11})

	Business										
	V_1	V_2	V_3	V_4	V_5	V_6	V_7	V_8	V_9	V_{10}	V_{11}
V_1	-	0.775	0.727	0.716	0.705	0.714	0.739	0.747	0.732	0.691	0.654
V_2	0.775	-	0.739	0.752	0.708	0.731	0.762	0.768	0.747	0.706	0.657
V_3	0.727	0.739	-	0.730	0.747	0.728	0.756	0.759	0.741	0.743	0.620
V_4	0.716	0.752	0.730	-	0.737	0.755	0.726	0.743	0.727	0.712	0.659
V_5	0.705	0.708	0.747	0.737	-	0.713	0.718	0.729	0.711	0.726	0.612
V_6	0.714	0.731	0.728	0.755	0.713	-	0.720	0.741	0.768	0.699	0.692
V_7	0.739	0.762	0.756	0.726	0.718	0.720	-	0.785	0.739	0.715	0.633
V_8	0.747	0.768	0.759	0.743	0.729	0.741	0.785	-	0.767	0.735	0.635
V_9	0.732	0.747	0.741	0.727	0.711	0.768	0.739	0.767	-	0.721	0.667
V_{10}	0.691	0.706	0.743	0.712	0.726	0.699	0.715	0.735	0.721	-	0.566
V_{11}	0.654	0.657	0.620	0.659	0.612	0.692	0.633	0.635	0.667	0.566	-

Table C.2 The 15 FCM simulations of the V_{11} values for the six observed cases in the Business sector. The upper-triangular FCM weighted matrix is derived from the distance-based method (Table C.1)

	Output of V11					
Scenarios	**Strength**	**Closeness**	**Between-ness**	**Mean**	**Random**	**Reverse_ strength**
V2	0.998	0.998	0.998	0.998	0.998	0.998
V7	0.947	0.947	0.947	0.947	0.947	0.947
V4	0.992	0.992	0.992	0.992	0.992	0.992
V8	0.905	0.905	0.905	0.905	0.905	0.905
V9	0.841	0.841	0.841	0.841	0.841	0.841
V2-V7	0.998	0.998	0.998	0.998	0.998	0.998
V2-V4	0.998	0.998	0.998	0.998	0.998	0.998
V2-V8	0.998	0.998	0.998	0.998	0.998	0.998
V7-V4	0.993	0.993	0.993	0.993	0.993	0.993
V7-V8	0.963	0.963	0.963	0.963	0.963	0.963
V2-V7-V4	0.998	0.998	0.998	0.998	0.998	0.998
V2-V7-V8	0.998	0.998	0.998	0.998	0.998	0.998
V7-V4-V8	0.993	0.993	0.993	0.993	0.993	0.993
V2-V3-V7	0.999	0.999	0.999	0.999	0.999	0.999
ALL Vi	0.999	0.999	0.999	0.999	0.999	0.999

The manufacturer's authorised representative in the EU is Springer Nature Customer Service Centre GmbH, Europaplatz 3, 69115 Heidelberg, Germany. If you have any concerns regarding our products, please contact ProductSafety@springernature.com

Printed and bound by CPI Group (UK) Ltd, Croydon, CR0 4YY

30/09/2026

02225515-0016